BIG
Data
Revolution

WHAT FARMERS, DOCTORS AND INSURANCE AGENTS TEACH US ABOUT DISCOVERING BIG DATA PATTERNS

Rob Thomas
Patrick McSharry

WILEY

This edition first published 2015
© 2015 John Wiley and Sons, Ltd.

Registered office
John Wiley & Sons Ltd, The Atrium, Southern Gate, Chichester, West Sussex,
PO19 8SQ, United Kingdom

For details of our global editorial offices, for customer services and for
information about how to apply for permission to reuse the copyright material
in this book please see our website at www.wiley.com.

A catalogue record for this book is available from the British Library.

ISBN 978-1-118-94371-7 (paperback); 978-1-118-94373-1 (ePDF);
978-1-118-94372-4 (ePub)

Set in 9.5/11.5 MinionPro-Regular by SPS/TCS

Printed in U.S. by Bind Rite Robbinsville

BIG
Data
Revolution

This book is for those who are willing to lead, in any endeavor. Most of all, this book is for Kristin, Will, Abby, and Sam. And, a special thanks to my big sister.

— Rob Thomas

To my parents Agnes and Patrick, wife Emmeline, and children Isolde, Theodore and Caspian.

— Patrick McSharry

About the Authors

Rob Thomas is Vice President of Product Development for Big Data and Information Management in the IBM Software Group. With over 15 years in the technology industry, Mr. Thomas has had the opportunity to consult to a variety of global businesses. He has experience in business and operational strategy, high technology, acquisitions and divestitures, manufacturing operations, and product design and development.

Mr. Thomas is currently responsible for product development and engineering for IBM's Big Data and Information Management product line. As Vice President of Business Development in IBM Software, Mr. Thomas led the acquisition of Netezza and Vivisimo, both leaders in the data era.

Mr. Thomas has extensive international experience, leading IBM's high technology services business in Asia Pacific, while living in Tokyo, Japan. Prior to that, he was a Partner in IBM's consulting business.

Mr. Thomas graduated from Vanderbilt University with a BA in Economics. He earned his Masters in Business Administration from the University of Florida. Mr. Thomas publishes regularly on his blog (http://www.robdthomas.com) and has an active following on Twitter (@robdthomas). He is an avid golfer, reader, and exercise enthusiast. He lives in New Canaan, Connecticut with his wife (Kristin) and three children (Will, Abby, and Sam).

Most of what he has learned in his life came from his parents, his wife, and his two sisters.

Patrick McSharry is a Senior Research Fellow at the Smith School of Enterprise and the Environment, Faculty Member of the Oxford Man Institute of Quantitative Finance at Oxford University, Visiting Professor at the Department of Electrical and Computer Engineering, Carnegie Mellon University, Fellow of the Royal Statistical Society and Senior Member of the IEEE. He takes a multidisciplinary approach to developing quantitative techniques for data science, decision-making, and risk management. His research focuses on big data, forecasting, predictive analytics, machine learning, and the analysis of human behavior. He has published over 90 peer-reviewed papers, participated in knowledge exchange programs and consults for national and international government agencies and the insurance, finance, energy, telecoms, environment, and healthcare sectors. Patrick received a first class honours BA in Theoretical Physics and an MSc in Engineering from Trinity College Dublin and a DPhil in Mathematics from Oxford University.

About the Technical Editor

Carin Anderson is a freelance technical editor. She has edited, compiled, and written numerous grants and proposals over the last decade and a half. Carin developed a mobile application company, creating multi-user gaming platforms. She also co-founded an informational website targeting families with young children.

In her spare time, she enjoys spending time with her family and friends, running and reading.

PUBLISHER'S ACKNOWLEDGEMENTS

Some of the people who helped bring this book to market include the following:

Editorial and Production
VP Consumer and Technology Publishing Director: Michelle Leete
Associate Director–Book Content Management: Martin Tribe
Professional Technology & Strategy Director: Barry Pruett
Commissioning Editor: Ellie Scott
Development Editor: Tom Dinse
Copy Editor: Laura Miller
Technical Editor: Carin Anderson

Marketing
Associate Marketing Director: Louise Breinholt
Marketing Manager: Lorna Mein
Marketing Executive: Polly Thomas

Composition Services
Proofreader: Wordsmith Editorial
Indexer: Potomac Indexing, LLC

Contents

Prologue

BERKELEY, 1930s

George Dantzig sat in his dorm room, contemplating the next 24 hours and what it would mean for his future. He came to the University of California, Berkeley, with many aspirations, but as often happens, life got in the way, and his best laid plans turned into dreams for another day. As he gazed over the building immediately in the foreground, he could see Sather Tower on Berkeley's campus, known for resembling Campanile di San Marco in Venice. George reassured himself that one of his major goals was still in his grasp; he could still earn a position on the faculty, providing an opportunity to teach the next group of eager students.

It was 3 p.m., and George had until 8 a.m. the next morning to prepare for what would become his defining exam at Berkeley. A passing grade virtually guaranteed his spot on the faculty. Anything less than his best, and his future would be once again uncertain. This was the kind of motivator that got him to reopen the books and apply himself throughout the night. The last time George looked up from his book, he saw 3 a.m. on the clock and decided he should get some rest.

As the sunrise slowly emanated around his room, George opened one eye and then the other, immediately wondering why he had not heard his alarm yet. He figured it must be an exceptionally clear day for that type of light to be coming through his eyelids before the 7:15 a.m. alarm that he set. Suddenly, George felt like something was not right, sat straight up in bed, grabbed his glasses, and looked at the clock: 8:30 a.m. The exam began 30 minutes ago! George quickly pulled on his pants and dashed for the door.

George sprinted into the exam hall, where the professor greeted him with a surprised look. Obviously, the professor concluded that George must be in the hospital or perhaps even dead to have missed the start of the exam. George, in a rushed voice, explained the situation as his professor handed him the exam. He also noted, "George, there are three additional problems that I have written on the board, once you complete the questions on the exam paper."

George, without any minutes to waste, sat in the front row and quickly started working through the questions. The exam was set for three hours, so when George arrived at 8:50 a.m., many of the students were nearly halfway through with the questions. Two hours later, as the clock approached 11 a.m., George finished the last question on the paper exam and shifted his attention to the three questions on the board. George was the only student left in the hall and clearly he would not have a chance to finish. He sheepishly walked up to his professor, re-explained the situation, and apologized that he did not have time to answer the questions on the board. In an unexpected act, his professor offered to let George have until midnight to try to complete the additional questions, and George excitedly ran back to his dorm room.

It was now 3 p.m., over 24 hours since he reflected on his future in his dorm room. George made progress on one of the questions but decided to give up on the others. He spent the next eight hours grinding on the first question, feeling confident that he had conquered the problem, and set out across campus to turn in his single answer. His disappointment was obvious in his posture — while he felt a great sense of accomplishment on one question, he knew that providing only one answer out of three would not be sufficient. George slid the paper under his professor's door, grabbed a small bite at the campus cafeteria, and collapsed into his bed at 1 a.m.

George was awakened by the shrill sound of his phone at 7 a.m., and he heard his professor on the line, "George, I can't believe it. You actually solved one of the equations on the board! This is truly an historic day." George, confused by what he was hearing, asked his professor to explain his amazement. His professor replied, "George, when I handed out the test, I told the class that I wrote three unsolvable equations on the board. I expected anyone with extra time to play around with them, but they weren't actually part of the test. You accomplished something that the rest of us KNEW was impossible!"

It's amazing what we can accomplish when we are not encumbered by what we believe is possible. It turns out that George had solved an algorithm around linear programming, which eventually became the simplex algorithm, the heart of Microsoft Excel's SOLVER function. If George had arrived on time to the exam hall and heard his professor tell the class to try to solve the unsolvable problems, he probably would have never accomplished this feat. *He was not limited by what the world felt was impossible.*

PATTERN RECOGNITION

In most great detective stories, the author typically uses a signature act or prop. The robber or murderer, either intentionally or unintentionally, leaves a signature mark at the scene of a crime. The detectives in pursuit do not notice it at first but eventually come to see a pattern. Often, it's about observing a non-obvious relationship between two data points, which leads to the recognition of a pattern. The conclusion: Insight can be found anywhere and hides in patterns. But are patterns valuable only to Sherlock Holmes and his contemporaries?

Patterns exist everywhere we look. However, it is the actual recognition of the pattern that can truly impact the world in which we live. When studying highly successful people, whether they're athletes or successful business professionals, Malcolm Gladwell argues in his best-selling book *Outliers* (Little, Brown and Company) that by studying the patterns of these individuals, we begin to understand that the environment from which they come from directly correlates to the amount of success they will have in life. By looking at hidden patterns of these individuals' upbringings, the month they were born, or the culture in which they were raised, we can predict whether they will reach their full potential or perhaps fall short.

The same can be said for data. By capturing and analyzing data, we can find patterns that will ultimately impact the future of industries and businesses. Often, observing a non-obvious relationship between two data points leads to the recognition of a pattern. The conclusion? Insight can be found anywhere and hides in patterns.

NELSON PELTZ

Nelson Peltz's coffee mug on his desk reads Sales Up . . . Expenses Down on one side, with Cash Is King on the other. These mantras helped him to build an $8.5-billion partnership focused on activist investing, one of the largest such partnerships in existence today. However, this was not handed to him; Peltz built this partnership through his ability to see, understand, and apply patterns across a variety of businesses.

Born in 1942 in Brooklyn, New York, Peltz worked his way through early school years and eventually decided to attend the Wharton School at the University of Pennsylvania. He dropped out a couple years later, in 1963, and set off to be a ski instructor. When that did not work out, he returned home to drive a delivery truck for his grandfather's company, A. Peltz & Sons. Eventually, he was given the reins to the company (Flagstaff) and grew it into a publicly held company. While Nelson did not finish a formal education, his on-the-job learning taught him all he would need to know later in his life.

In the 1980s, Nelson reunited with a former business partner from Flagstaff, Peter May, and they went on a hunting expedition: looking for companies to acquire, grow, and eventually sell. His first marked success was the sale of Triangle Industries to Pechiney in 1988. Nelson began to notice the value of spotting patterns in the operations of the companies that he acquired or considered acquiring.

Eventually, Trian Fund Management was founded, and still exists today. He became a force advocating for change and was an activist investor before the term was en vogue. But how did he do it?

Peltz believes that most activist investors and private equity companies focus on financial engineering. While that can make a difference in many cases, it gets you only so far. Very few individuals possess the insight to improve business operations, which is where you can drive substantial returns. And, in Peltz's view, improving business operations requires identifying and understanding patterns, and then acting on them.

To drive meaningful insight out of patterns, Peltz consciously (or unconsciously) focused much of his investment in two sectors: consumer packaged goods and food. With a focused competence, he can better compare and contrast patterns of performance. He believes in constantly assessing companies in his area of competence and looking at their strengths and weaknesses against the best performers in the industry. A sampling of the patterns that he looks for include:

- **Ratios:** Percent of sales spent on marketing.
- **Overhead:** Growth in overhead versus growth in sales.
- **Rebates and allowances:** Deals and allowances paid to retailers.
- **Brand:** He favors companies that have strong brands, but also have ratios that are out of line with the best performers.

Peltz believes that it's a more efficient use of capital to revitalize previously great brands than to try to build a new one. This is why he began to take an active interest in Heinz and Wendy's.

Heinz

When Peltz began analyzing H.J. Heinz Company in 2005, it was an immediate fit to his pattern-focused investing style. He saw a company with brand value and strong free cash flow, yet the total shareholder returns trailed the S&P 500, the large-cap food index, and the mid-cap food index. But why?

He observed that Heinz's Selling, General, and Administrative (SG&A) expenses, as a percent of revenue, was dramatically higher than the best comparable performers. The advertising investment as a percent of revenue was also out of line. Next, he noticed that the rebates and allowances being paid to retailers were much higher than the amounts paid by other organizations in its peer group. In his mind, that money could be put into marketing and product innovation, instead of lining the pockets of the retail channel. He noticed that plant efficiency metrics also trailed the best performers.

Lastly, he highlighted the fact that all of Heinz's businesses were operating at margins in excess of the company average, indicating that the overhead at headquarters was crippling the business. All the patterns that he had learned to watch for over time were present at Heinz.

Wendy's

Wendy's, like Heinz, demonstrated a pattern of underperformance versus its peer group. In the case of Wendy's, through a ratio analysis, Peltz could see that margins were unacceptably low (10 percentage points below its peer group) and driven by excessive overhead and operating costs. Next, Wendy's lost its focus on brand strength, as it had diversified into other food categories like Tim Hortons Cafe and Bake Shop in Canada. Different company, same patterns.

While it seems simple to observe after the fact, these were new insights, previously unnoticed at the time.

Peltz has gone on to advocate for similar change in the likes of Cadbury-Schweppes, Kraft Foods, Snapple Beverage Corp., PepsiCo, and many others. All these companies fell within his core competence, all assessed against the same set of patterns, and each one was driving value for Peltz, his investors, and the investors in those companies. In the last few years, the assets under management at Trian increased from $3.7 billion in 2012, to $6.3 billion in 2013, to $8.5 billion in the most recent report.

The power of being able to identify, understand, and execute upon patterns of success is critical in the pursuit of distinguishing oneself or an enterprise.

COMMITTING TO ONE PERCENT

Dave Brailsford basked in the glow of the Olympics. The Great Britain cycling team just completed their participation at the 2012 Olympics in London, England, winning 70 percent of the medals in men's cycling. Reporters probed with aggressive questioning, wanting to understand the silver bullet that led to this success. The irony: There was no silver bullet. In fact, it was the opposite of a silver bullet.

When the Great Britain cycling team, Team Sky, hired coach Dave Brailsford in 2010, the country had never won a Tour De France. In fact, the history of the sport in the country was filled with errors, mishaps, and minimal success. Historically, the team chased fads of success: new equipment, new uniforms, new techniques. But nothing changed the trajectory. Then, Dave Brailsford arrived.

Dave Brailsford fanatically talks about the aggregation of marginal gains. This concept means that by marginally improving each and every aspect of a process, the aggregation of those small gains will lead to large improvements. Brailsford's goal was simple: one percent. He sought a one-percent improvement in every aspect of the cycling team.

Setting out to improve all aspects of a cycling team, the obvious places to start are in areas like nutrition, bike performance, and physical conditioning. After all, improving every meal by one percent promised a path to continued improvement. However, for Brailsford, those enhancements merely scratched the surface. He set out to improve *every* aspect by one percent. Not only sports massage, but the gels used for sports massage. Not only the bikes, but the grips on the bikes and, more specifically, the tackiness of the grips. He studied not only the physical conditioning, but also the sleep habits and, more specifically, the pillows used. He focused on every aspect: one-percent improvement. It's that simple.

In 2012, a short two years after Brailsford joined the team, Great Britain won its first Tour De France. Shortly thereafter, the triumph at the Olympics in London occurred. The aggregation of the one-percent gains created superior outcomes.

THE BIG DATA REVOLUTION

The big data revolution is about accomplishing feats with data that no one believes is possible. The leaders of the big data revolution will embody three characteristics:

- The ability to suspend disbelief of what is possible, and to create their own definition of possible
- An inherent knowledge of pattern recognition and the insight to apply patterns from one industry or dimension to another that may be seemingly unrelated
- Commitment to one-percent improvement in every aspect related to data

Combining these seemingly different characteristics and applying them creates possibilities previously undetected. By empowering ourselves with data and believing we can discover the undiscovered, we can launch businesses and industries to new levels. Throughout this book, we will look at different industries and how they use data to make their impossible possible.

This revolution is about finding your possible.

Introduction

STORYTELLING

Stories make for powerful communication, and this book is a compilation of the stories, patterns, and methods that we have seen over the last decade, since we've been focused on this emerging phenomenon: Big Data. The stories are based on true events but do not always include actual names, events, or circumstances.

These stories are meant to illustrate the challenges and possibilities present with the advent of big data, based on what we have witnessed. Our belief is that such stories provide the best way to learn about how other business leaders both responded to external change and in some cases caused disruptive change within their respective industries. We hope that they will provide a source of inspiration, courage, and know-how, so that you can embrace big data as a means of inciting a revolution within your organization.

OBJECTIVE

The target audience of this book ranges from the entrepreneur, to management in established enterprises, to those who are merely curious about the implications of big data in their own lives. This book is intended to provide ammunition for breaking down the barriers that often exist between those who manage data and those who manage people. Data is the new intellectual property. It can be harnessed for advantage or ignored at peril.

For those who would prefer to remain working in silos, where data analysis and decision-making is divorced, this book will make for uncomfortable reading. Organizations that do not manage to utilize their data assets will eventually become extinct. The challenge of improving connectivity between data and behavior, and between machine and human, will require dedicated effort in terms of building human capacity and financial resources. In addition, patience is required while organizations make this transition. For some organizations, it may be necessary to initiate external activities or form partnerships in order to adequately assess the potential value of big data.

Unfortunately, many have a vested interest in resisting the data revolution due to their fears about the impact it will have on their own professions. It is likely that such resistance will be futile and that those who actively embrace the oncoming disruptive change will benefit most from the opportunities offered. Estimates by C.B. Frey and M.A. Osborne (*"The Future of Employment"* [Oxford University, 2013]) suggest that almost half of existing jobs will be at risk of automation due to the technological developments that are currently taking place because of big data and the application of machine-learning approaches. While many routine tasks are already being computerized and automated, recent scientific advances suggest that it will be possible to automate an increasing amount of non-routine cognitive tasks, such as accountancy, legal work, technical writing, and many other white-collar occupations.

We argue throughout the book that in order to develop strategies for managing organizations in a knowledge-based society, it will be necessary to grab hold of data opportunities before agile, data-savvy competitors pass you by. The transition needed is complex and involves changes to technology, processes, and human behavior. There is no doubt that surviving such disruptive change will be a substantial challenge for many. While there is no magic formula to fit all types and sizes of organizations, we offer a strategy in the form of a roadmap that can be easily tailored to suit each individual organization.

Our approach to creating a roadmap for success in the Data era is to first explain how other organizations have changed and outsmarted their competitors. The first part of the book presents nine stores about innovation using data. The second part of the book gathers together the key patterns that we have observed through the stories presented. By distilling these patterns, we offer a broad understanding about how data is being leveraged across a range of different industries. Finally, in the third part of the book, we aim to inspire business managers to lead the revolution by offering a methodology for operationalizing big-data approaches that can be adapted for different industries.

OUTLINE

PART I "THE REVOLUTION STARTS NOW: 9 INDUSTRIES TRANSFORMING WITH DATA"

Part I shares the main stories that will be used to illustrate patterns and concepts throughout the book. While we had hundreds of stories to choose from based on our experiences, we chose the nine that we think are the most compelling and engaging, and that bring out the best illustration of the patterns that we have seen. The stories can be read in any order, so we suggest you pick the one that sounds the most interesting, and then work your way through the others:

- **Chapter 1: Transforming Farms with Data.** Rob starts with a quick look at the history of farming and the evolution of technology in agriculture. The chapter delves into limitations through the years and how they have been overcome. The majority of the chapter explores how data will continue to transform this seemingly analog industry. As you will see, many players in this industry have not yet awakened to the impact of data, and they are quickly being passed by, perhaps without even noticing.

- **Chapter 2: Why Doctors Will Have Math Degrees.** Historical approaches to medicine, treatments, and wellness are not relevant in the data era. The chapter highlights how many decisions in the medical field today are based on opinions instead of facts, and how this leads to suboptimal outcomes. We also showcase a new set of leaders in medicine who are disrupting traditional industry practices through the use of data. The key implication is that the role and skills of doctors will change in the Data era.

- **Chapter 3: Revolutionizing Insurance: Why Actuaries Will Become Data Scientists.** The insurance industry is undergoing a fundamental shift based on better collection, access, and usage of data. Underwriting and actuarial services, which are largely about forecasting what might happen, will take a backseat in a world where you can monitor what is actually happening and price accordingly. New business models are emerging, which is disrupting the traditional skills and tools needed to win in insurance.

- **Chapter 4: Personalizing Retail and Fashion.** Data can turn traditional segment-based retailing into a more personal approach: thousands of individual customers, instead of thousands of customers. There is a timeless quote in retail stating, "I know that only half of my marketing is effective. The problem is that I don't know which half is working." Transforming retail, however, is more than just using data to better target clients. It's about using data to transform the role of a retailer and truly serve a customer of one.

- **Chapter 5: Transforming Customer Relationships with Data.** Data will increase the intimacy between the firm and the customer. By improving the collection, relevance, and utilization of data about customers, firms will be able to maximize customer satisfaction by processing data about individuals in real-time. Rather than responding to issues and problems, data about the locations and preferences of individuals will allow organizations to offer services and solutions to improve their personal experiences and identify challenges ahead of time.

- **Chapter 6: Intelligent Machines.** Rob starts with a story about visiting Denmark and his exploration of the wind turbine business. The chapter goes on to discuss how previously unconnected machines are coming to life through connectivity and data. Bringing life to machines may seem futuristic, but there are already many developments that are facilitating the production of machine-readable data that is helping to increase intelligence. The Internet of Things describes the network of such machines and their ability to share information. The Industrial Internet heralded by General Electric, intelligent wind turbines, the potential of drones, and Tesla's Vehicle Management System are all examples of how networks of data will revolutionize our world.

- **Chapter 7: Government and Society.** Closing the loop between people and government using data has considerable potential. Whereas elections, referendums, and opinion surveys cost substantial amounts of money, social media offers a means of monitoring public opinion, assessing perceptions, and testing and fine-tuning public policy. At the same time, privacy risk is now a major concern in many countries and is delaying data open-access initiatives. Finding a reliable way to address these risks through anonymization techniques without degrading the quality of the data will be a challenge. This chapter also explores the potential rewards of using big data for public-private partnerships for delivering socio-economic benefits.

- **Chapter 8: Corporate Sustainability.** Connecting people through the Internet and social media has increased awareness about the global supply chain behind many of the services and products that we consume on a daily basis. Faced with the collective responsibility for ensuring sustainable practices, many firms are now seeking to become leaders within their industry and are also reaping the benefits of moving first. Having the confidence to design and implement a corporate sustainability strategy requires the capability to assess the risks associated with future scenarios. Agitation for change is coming from those that have the mandate to make long-term decisions.

- **Chapter 9: Weather and Energy.** Forecasting the weather has been and will continue to be a challenge, despite the many scientific advances that have been made in terms of data, models, and techniques. Nevertheless, weather forecasting serves to illustrate how human behavior relates to the task of generating and responding to future scenarios. The close relationship between weather and energy shows how big data will be used to operate power systems when substantial amounts of variable renewable energy are integrated. Although introducing many changes, the combination of better data and technology innovation will help to balance supply and demand and keep the lights on.

PART II "LEARNING FROM PATTERNS IN BIG DATA"

The second part of the book distills the nine stories in Part I into a set of discrete patterns. We explore the concept of pattern recognition, how it can be applied in a multitude of settings, and the implications for the Data era. We close this part of the book with a detailed discussion of the 54 patterns in big data that we have observed:

- **Chapter 10: Pattern Recognition.** Identifying patterns is usually the first step to constructing systems for forecasting, classifying, or simply segmenting customers. The ability to identify patterns in big data, and to also assign significance levels to these patterns, will be extremely important in the Data era. Collecting relevant data, cleansing this data, extracting appropriate features, classifying the data, and evaluating confidence is all part of the process. From understanding Bayes theorem to distinguishing species in a Tokyo fish market, pattern recognition algorithms vary from human intuition to sophisticated machine-learning algorithms.

- **Chapter 11: Why Patterns in Big Data Have Emerged.** Emerging patterns in applications of big data in different organizations can help illustrate the potential of this revolution. There are three prominent approaches to building a business model in the Data era. In some cases, data provides a competitive edge by moving before the crowd. In others, it improves the existing products or services. Finally, data can become the product itself by recognizing the need for firms to obtain access to particular datasets.

- **Chapter 12: Patterns in Big Data.** Identifying patterns in big data can help anyone prepare their organization for the Data era. A series of 18 data factors are described, based on the stories from Part I of the book. These factors are then further decomposed into 54 big-data patterns, with the aim of representing the best practices of a range of leading-edge firms in the Data era.

PART III "LEADING THE REVOLUTION"

The third part of the book focuses on how to create a big-data revolution in your own organization. You need to develop an appreciation for the lessons learned from the individual stories in the first part of the book, coupled with a bias for action. We recommend thoughtful consideration about how the patterns that have been extracted in Part II of the book apply (or don't) to your organization. The substantial challenge facing each individual business leader is to determine the steps necessary for operationalizing the required changes in as short a time as possible:

- **Chapter 13: The Data Opportunity.** Focusing on opportunity is important to encourage behavior change. Similarities between the processes of finding, refining, and adding value for two commodities, oil and data, help to illustrate how value will be created in the future. Empirical support for the benefits of a data focused strategy first come from a Bain report published in 2013. Early adopters of big data are twice as likely to be in the top quartile of financial performance within their industries.

- **Chapter 14: Porsche.** Aiming for perfection is key to success. Rob begins with a story set in Italy, and expands into a story about Ferdinand Porsche, the founder of Porsche. The story serves to demonstrate why innovation, adaptability, and perseverance are responsible for the success of this impressive sports-car business. Porsche's fit-for-purpose approach ensures that their vehicles are designed to deliver to many different client needs while optimizing performance, quality, and value. A similar approach, fit-for-purpose, will be taken by the leaders of the Data era.

- **Chapter 15: Puma.** Striving to be the best is one of the hallmarks of a successful organization. This story of the aggressive battle between Puma and Adidas is rooted in sibling rivalry and went well beyond business competition. Jochen Zeitz took over the reigns of Puma in 1993 and delivered a 4,000-percent increase in Puma's share price over the next 13 years. Recognized as a great business leader, Zeitz has used big-data analytics to create an environmental profit-and-loss accounting system and is now advocating its use in other firms.

- **Chapter 16: A Methodology for Applying Big-Data Patterns.** This methodology, divided into seven specific steps, provides a recommended approach for applying big-data patterns in any organization. While the methodology is intended to be applied sequentially, some organizations may have already completed (or at least started) some of the steps. The methodology should be used as a roadmap, as opposed to a destination: Use the parts that you need.

- **Chapter 17: Big-Data Architecture.** As an organization works through the methodology in Chapter 16, the thinking will often turn to execution. Chapter 17 takes the first steps towards execution, sharing the landscape of a big data reference architecture. The focus is on Business View and Logical View reference architectures.

- **Chapter 18: Business View Reference Architecture.** This chapter dives into the components of a Business View reference architecture. We introduce you to a fictional retailer, Men's Trunk, which illustrates a journey into the Data era. The Business View reference architecture includes the Answer Fabric, Data Virtualization, Data Engines, Management, Data Governance, and User Interface/Applications components.

- **Chapter 19: Logical View Reference Architecture.** Once a company understands its reference architecture from a Business View, applying it for impact requires a Logical View of the architecture needed. A more granular view into how to execute for the Data era, the Logical View architecture focuses on five components: Data Ingest, Analytics, Information Insight, Operational Data, and Governance.
- **Chapter 20: The Architecture of the Future.** The book finishes with a detailed look at the journey of Men's Trunk, the fictional retailer introduced in Chapter 18. We track Men's Trunk through their application of the methodology and the implementation of reference architectures.

STORYTELLING (CONTINUED)

We are storytellers due to our belief that it is the most effective way to communicate for impact. Each of the parts and chapters in this book are interlaced with stories and illustrations, intended to capture interest and to create enduring images in the mind of the reader. As you navigate through the Data era, we hope these stories serve not only as lessons and guiding points, but also as inspiration. Change is difficult, and inspiration is often needed to keep going onward.

This book, like everything in the Data era, will not stand still. The tools and the environment will change, necessitating change in approaches and techniques. You can keep abreast of any changes and view relevant content and interviews on the book's website, www.bigdatarevolutionbook.com.

THE REVOLUTION STARTS NOW: 9 INDUSTRIES TRANSFORMING WITH DATA

Chapter 7: Government and Society

Chapter 8: Corporate Sustainability

Chapter 9: Weather and Energy

1

TRANSFORMING
FARMS WITH DATA

CALIFORNIA, 2013

AS THE WHEELS came down on my cross-country flight, I prepared for our landing at San Francisco International Airport (SFO). Looking out the window, I could see the sprawl of Silicon Valley, the East Bay, and in the distance, the San Francisco skyline. It is hard to believe that I was here to explore agriculture in 2013, given that what I could see from the plane was mostly concrete, highways, and heavy construction.

Not too many miles away from SFO, I began to wind through the tight curves of back roads, making my way to the headquarters of a major agricultural producer. While I had never visited this company before, I had the opportunity to sit down with the executive team to explore the topic of big data in farming and agriculture.

I embraced the calm and serene scene, a far cry from the vibrancy of San Francisco and the rush of Silicon Valley. As we entered a conference room, the discussion turned to produce, as I asked, "Why is it that the strawberries that I bought last week taste so much better than the ones I bought the week before?" While I posed the question as a conversation starter, it became the crux of our discussion.

It seems that quality — and, more specifically, consistency of quality — is the foremost issue on the mind of major producers. I asked about the exquisite quality of produce in Japan. The executive team quickly noted that Japan achieves quality at the price of waste. Said another way, they keep only 10 percent of what a grower provides. This clarified the point in my mind that quality, consistency of quality, and eliminating waste create the three sides of a balanced triangle.

The conversation that followed revealed one significant consensus in the room: Weather *alone* impacts crop production and the consistency of crops. And since no one in the room knew how to change the weather, they believed that this was the way things would always be. I realized that by blaming the weather this team believed their future did not belong in their own hands but was controlled by the luck, or the misfortune, of each passing season.

BRIEF HISTORY OF FARMING

The evolution of farming in the developed world provides context to much of the conventional wisdom about farming that exists today. Dating back to the 1700s, farming has been defined by four eras:

- **1700s (Subsistence Farming):** Farmers produced the minimum amount of food necessary to feed their families and have some in reserve for the cold winter months.
- **1800s (Farming for Profit):** This era marked the transition from subsistence farming to for-profit farming. This is when the widespread use of barns began, for the purpose of storing tools, crops, and related equipment. These were called pioneer farms.
- **Early 1900s (Power Farming):** At this time, the "power" came in the form of 1,800-pound horses. The farmers used animals for plowing, planting, and transporting crops. The use of animal labor drove the first significant increase in crop productivity.
- **Mid- to late 1900s (Machine Farming):** Sparked by the Industrial Revolution, this era's farmers relied on the automation of many of the tasks formerly done by hand or animal. The addition of machinery created tremendous gains in productivity and quality.

Each era represented a significant step forward, based on the introduction of new and tangible innovations: barns, tools, horses, or machines. The progress was *physical* in nature, as you could easily see the change happening on the farms. In each era, production and productivity increased, with the most significant increases in the latter part of the 20th century.

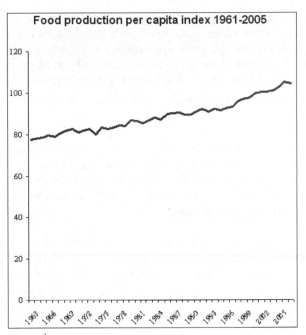

Farm productivity over time

Through these stages, farming became more productive, but not necessarily more intelligent.

THE DATA ERA

The current era of farming is being driven by the application of data. It is less understood than previous eras because it is not necessarily *physical* in nature. It's easy to look at a horse and understand quickly how it can make farm labor easier, but understanding how to use geospatial information is a different proposition. The advantage is driven by intangibles: knowledge, insight, decision making. Ultimately, data is the fertilizer for each of those intangibles.

A simple understanding of how a crop grows can aid in understanding the impact of data on farms. The basic idea is that a plant needs sunlight, nutrients from the soil, and water to grow into a healthy plant, through a process called photosynthesis. Healthy plants must keep cool through a process called transpiration (similar to how a human sweats when physically stressed). But, if a plant lacks the nutrients or conditioning to transpire, then its functions will start to break down, which leads to damage. Using data to improve farming is fundamentally about having the ability to monitor, control, and if necessary, alter these processes.

Today, according to the Environmental Protection Agency, there are 2.2 million farms in the United States and many more outside of the U.S. The average farm spends $110,000 per year on pest control, fertilizer, and related items to drive yield. The prescient way to improve profit and harvest yields across a vast territory requires better collection, use, and application of data.

POTATO FARMING

Potato farming can be exceedingly difficult, especially when attempted at a large scale with the goal of near perfect quality. The problem with potato farming is that the crop you are interested in is underground; therefore, producing a high-quality and high-yield potato crop depends on agronomic management during the growing process.

At the Talking Data South West Conference in 2013, Dr. Robert Allen, a Senior GIS Analyst at Landmark Information Group, highlighted the importance of data in potato farming, in his talk titled, "Using Smartphones to Improve Agronomic Decision Making in Potato Crops." Dr. Allen makes the case that leveraging data that describes the growth and the maturation of a crop during the growing season is instrumental to a successful yield. Continuous insight, delivered throughout the growing season, may have a material impact on the productivity of a crop.

One of the key variables required for yield prediction, and needed to manage irrigation, is groundcover. Groundcover, which calculates the percentage of ground covered by green leaf, provides critical input in the agronomic management of potato crops. Measuring groundcover is not as simple as pulling out a measuring tape; it requires capturing imagery of potato crops and large-scale collection of data related to the images (the water balance in soil, etc.), and the data must be put in the hands of farm managers and agronomists so that they can actually do something about what the data is telling them. The goal is not to collect data, but to act on it.

Dr. Allen describes four considerations in potato farming related to using data:

- **Time:** Data needs to be collected at regular intervals and decisions need to be made in near-real time.
- **Geography:** These tend to be large-scale operations (10,000 to 20,000 acres), with fields distributed over large areas.
- **Man power:** Data is often collected by farm field assistants (not scientists) and must be distributed because decision makers tend to be remote from the field.

- **Irrigation:** Irrigation, while very expensive, is a primary factor in the maturation of a potato crop. Utilizing data to optimize the use of irrigation can lead to a productive crop, at the lowest possible cost.

These considerations led to a data collection and analysis solution called CanopyCheck. While it requires only a download from Apple's App Store, it provides a rich data experience to compare groundcover and other related data to optimize the quality and yield of a potato crop.

The Landmark Information Group describes CanopyCheck (`http://download.cnet.com/ios/landmark-information-group/3260-20_4-10094055.html`) as

> *This app is for potato growers, using the CanopyCheck ground-cover monitoring system, and captures accurate and reliable images of the potato crop which can be used to describe crop development. Each image is geo-located and labelled with farm and field information specified by the potato grower on the accompanying CanopyCheck website.*

Conventional wisdom states that growing potatoes is easy: They don't need sunlight, they do not need daily care, and by controlling the amount of water they receive, growing potatoes is a fairly simple process. However, as is often the case, conventional wisdom overlooks the art of the possible. In the case of potatoes, the application of data and agronomy can drive yield productivity up 30 to 50 percent, which is material in terms of the economics and the waste that is reduced.

PRECISION FARMING

Whether you strike up a conversation with a farmer in the 1800s, 1900s, or even in the early part of this century, they would highlight:

- Their growing strategy evolves each year.
- While the strategy evolves, their ability to execute improves each year, based on increased knowledge.

While this farming approach has been good enough for the better part of three centuries, the Data era ushers in the notion of precision farming. According to Tom Goddard, of the Conservation and Development Branch of Alberta Agriculture, Food and Rural Development, the key components of precision farming are:

- **Yield monitoring:** Track crop yield by time or distance, as well as distance and bushels per load, number of loads, and fields.

- **Yield mapping:** Global Positioning System (GPS) receivers, along with yield monitors, provide spatial coordinates, which can be used to map entire fields.

- **Variable-rate fertilizer:** Managing the application of a variety of fertilizer materials.

- **Weed mapping:** Mapping weeds using a computer connected to a GPS receiver while adjusting the planting strategy, as needed.

- **Variable spraying:** Once you know weed locations from weed mapping, spot control can be practiced.

- **Topography and boundaries:** Creating highly accurate topographic maps using a Differential Global Positioning System (DGPS). This data can be used to take action on yield maps.

- **Salinity mapping:** This is valuable in interpreting yield maps and weed maps, as well as tracking the salinity over a period of time.

- **Guidance systems:** Guidance systems, such as DGPS (accurate to a foot or less) are valuable for assessing fields.

- **Records and analyses:** Large data collection is necessary to store pertinent data assets, along with images and geospatial information. It is important that this information can be archived and retrieved for future use.

The extensive insight that can be gained by collecting each of these data points is potentially revolutionary. It evolves a process from instinctual to data-driven — which, as seen in the potato example, has a fundamental impact on yields and productivity.

The underlying assumption is that the tools and methodology for capturing farm data are available and utilized efficiently. This is a big assumption because many farms today are not set up to actively collect and capitalize on new data assets. Accordingly, the ability to capture farm data becomes the source of competitive advantage.

CAPTURING FARM DATA

It sounds easy. Collect data. Then use that data to deliver insights. But, for anyone who has been on a rural farm in the last decade, it is easier said than done. There are limitations that exist on many farms: lack of digital equipment, lack of skilled technology labor, poor distribution of electricity, and poorly defined processes. Because of these factors, each farmer must establish a new order of doing things to take advantage of the Data era.

The data landscape for farming consists of three primary inputs.

Data landscape for farming

- **Sensing equipment:** Mounted devices on machinery, in fields, or anywhere near crops could be designed to collect/stream data or to control the application of water, pesticides, etc. This could range from instrumented tractors for harvesting to devices to monitor crop transpiration. The evolution of machines to collect data on crops and soil has been dramatic. In the last decade alone, equipment has evolved from mechanical-only to a combination of mechanical and digital technology. This change has been expedited by early insights that even small adjustments in planting depth or spacing can have huge impact on yields. So, while today the sensing equipment is largely a digitized version of common farm machines, the future will see a marked advancement in machines. Drones, driverless tractors, and other innovations will become commonplace.

- **Global Positioning System (GPS):** GPS provides the ability to pinpoint location accuracy within one meter. While GPS first emerged for automobiles in the early 1990s in places like Japan, it has just now become common in all automobiles. Farming equipment, as you may expect, has been even a step further behind, with the wide use of GPS just accelerating in the last decade.

- **Geographic Information System (GIS):** GIS assesses changes in the environment, tracks the spread of disease, as well as understanding where soil is moist, eroded, or has experienced similar changes in condition. Once you know weed locations from weed mapping, spot control can be implemented. Topography and geology are important considerations in the practice of farming. Both are well accounted for with modern-day Geographic Information Systems.

By combining these three inputs, farmers will be able to accurately pinpoint machinery on their farms, send and receive data on their crops, and know which areas need immediate attention.

DEERE & COMPANY VERSUS MONSANTO

John Deere founded Deere & Company in 1836, when he moved to Grand Detour, Illinois to open a repair shop for farming tools. Deere eventually moved beyond tools and into the production of plows, which became a mainstay in the Farming for Profit and Power Farming eras. In 1848, Deere relocated to its still-current home in Moline, Illinois, and after his death in 1886, he passed the presidency of the company to Charles Deere.

Charles led the company into the 20th century, where the company pioneered the move to gasoline tractors, which became the defining product of not only the company, but of farming and agriculture in this time. The dominance of the company was ensured by continuous innovations in their tractors, innovation in their business model (a robust dealer network), and their defining image: John Deere green. As of 2010, the company employed 55,000 people and was operating in 30 countries worldwide. A shoe-in for continued dominance, right?

Monsanto, founded in 1901, took a bit longer to come into its defining moment. Moving into detergents and pesticides, Monsanto eventually became the pioneer in applying biotechnology to farming and agriculture. With biotechnology at its core, Monsanto applies data and insight to solve problems. Accordingly, Monsanto was a data-first company in its birth, which continued to drive its innovation and relevance. But sometimes, it takes time for an industry to catch up to its innovative leaders, and the first major evidence of how Monsanto would lead a change in the landscape was seen around 2010. That is when you see the fortunes of Deere & Company and Monsanto start to go in different directions.

Monsanto had one critical insight: Establishing data-driven planting advice could increase worldwide crop production by 30 percent, which would deliver an estimated $20-billion economic impact — all through the use and application of data. As Monsanto bet the company on the Data era, the stock market began to realize the value of the decision, leading to a period of substantial stock appreciation.

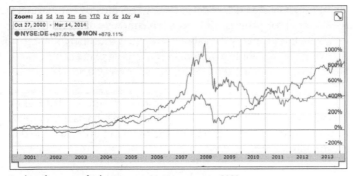

Zoom: 1d 5d 1m 3m 6m YTD 1y 5y 10y All
Oct 27, 2000 - Mar 14, 2014
●NYSE:DE +437.63% ●MON +879.11%

Stock performance of John Deere versus Monsanto since 2000

Data is disrupting farming, and we are starting to see that in the business performance of companies driving innovation in the industry. Gone are the days in which a better gasoline tractor will drive business performance. Instead, farmers demand data and analytics from their suppliers, as they know that data will drive productivity.

INTEGRATED FARMING SYSTEMS

Monsanto calls their approach to farming in the Data era, Integrated Farming Systems (IFS). Their platform provides farmers with field-by-field recommendations for their farm, including ways to increase yield, optimize inputs, and enhance sustainability. Listening to the data and making small adjustments to planting depth or the spacing between rows makes a vast difference in production. As Monsanto says, this is "Leveraging Science-Based Analytics to Drive a Step Change in Yield and Reduced Risk."

Monsanto's prescribed process for Integrated Farming Systems involves six steps:

1. **Data backbone:** Seed-by-environment testing to produce on-farm prescriptions
2. **Variable-rate fertility:** Adjusting prescriptions, based on conditions.
3. **Precision seeding:** Optimal spacing between rows
4. **Fertility and disease management:** Custom applications, as needed
5. **Yield monitor:** Delivering higher resolution data
6. **Breeding:** Increase data points collection to increase genetic gain

FieldScripts became the first commercial product to be made available as a component of Monsanto's overall IFS platform. FieldScripts provides accurate seeding prescriptions for each farmer and each field.

Monsanto, through its seed dealer network, engages directly with farmers to optimize two variables: planting and seeding technology data. The seeding technology, which is primarily data about seeding, is the differentiating factor. Applying that insight to a personalized planting plan enables Monsanto to deliver personalized prescriptions for every field.

FieldScripts, delivered via iPad, utilizes a custom application called Field-View. FieldView, deployed to farmers, while leveraging the data acquired throughout the years, equips farmers with the tools and insights needed to make adjustments for optimal yields.

Deere & Company and Monsanto both have bright futures. According to Jeremy Grantham, chief investment strategist of Grantham Mayo Van Otterloo (GMO), with the world's population forecasted to reach almost 10 billion by 2050, the current approach cannot sustainably feed the world's population. The demand presented by population growth creates an opportunity for all companies that service the industry. For the moment, Monsanto has leaped ahead in this new era of data-farming over the past five years, forcing Deere & Company to play catch-up.

DATA PREVAILS

Data is starting to prevail in agriculture. This is evident not only in the changing practices of farmers, but also in the ecosystem. New companies are being built, focused on exploiting the application of data.

THE CLIMATE CORPORATION

Monsanto's aggressive move into the Data era was perhaps punctuated in October 2013 with their announced acquisition of the Climate Corporation for $930 million. Why would a firm with its roots in fertilizers and pesticides spend nearly $1 billion on an information technology (IT) company? This aggressive acquisition demonstrates the evolution of the industry. "The Climate Corporation is focused on unlocking new value for the farm through data science," commented Hugh Grant, the chairman and chief executive officer for Monsanto. Founded in 2006, the Climate Corporation employs a team unlike any other in the agriculture industry. The team is composed of many former Google employees, along with other elite technology minds from the Silicon Valley scene. The tools they develop help farmers boost productivity, improve yields, and manage risks, all based on data.

At the heart of this acquisition lies the core belief that every farmer has an unrealized opportunity of around 50 bushels of crop (corn, potatoes, etc.)

in each of their fields. The key to unlocking these additional bushels lies in the data.

While the leaders of the past would provide better machines, Monsanto focuses on providing better data. By combining a variety of data sources (historical yield data, satellite imagery, information on soil/moisture, best practices around planting and fertility), this information equips the farmers with the information they need to drive productivity.

GROWSAFE SYSTEMS

GrowSafe Systems began studying cattle in 1990. This was not a group of former cattle hands, but a team of engineers and scientists who foresaw data science as playing a role in cattle raising. In 2013, the GrowSafe team won the Ingenious Award from the Information Technology Association of Canada for best innovation. This was the first time that this organization gave an innovation award to anyone in the world of cattle.

GrowSafe developed a proprietary way of collecting data through the use of sensors in water troughs and feedlots. With these sensors, they track every movement of cattle, including specifics about the cattle themselves: consumption, weight, movement, behaviors, and health. Each night, the data is collected and then compared against a larger corpus of historical data. The goal is to look for outliers. GrowSafe knows that the data reveals information that cattle farmers often cannot detect. This innovative approach enables farmers to prevent a disease before it begins.

FARM OF THE FUTURE

The mainstays of today's farms are people, fertilizers, irrigation, gas machines, trucks and carts for transport, and local knowledge. It is a craft, and typically the only person that can run a certain farm is the person that started it. This is why so many farms fold after the head of the operation retires. The success is in their hands — it's their craft.

Farms in 2020 will have a completely different feel from today's farms. In fact, they may be unrecognizable to a farmer of the early 21st century. Approaches that seem futuristic today will be common to all farmers in 2020:

- **Digital machines:** Digital machines, acting as sensors, will be the norm. The days of simple gas machines will be far in the past. In fact, by 2020, many farm machines will be battery- or solar-powered, with gas itself becoming a rarity. The digital machines will be much more than Internet-enabled tractors. There will be drones. Many drones. In fact, drones will become the most cost-effective and precise mechanism for managing many chores that farmers do via hand or by tractor today. With the sprawl of digital machines, device management will become the "cattle herding of the future," as all devices will have to be managed and maintained appropriately.

- **IT back office:** Every farm will have an information technology (IT) back office. Some will manage it themselves, while many will rely on a third party. The IT office will be responsible for the aforementioned device management, as well as remote monitoring and, ultimately, data-driven decision making. The IT back office will be the modern-day farm hand, responding to the farmers' every need and ultimately ensuring that everything operates as programmed.

- **Asset optimization:** With the sprawl of new devices and machines, asset optimization will be at the forefront. Maximizing the useful life of machines, optimizing location, and managing tasks (workloads) will be key inputs into determining the productivity of a farm.

- **Preventative maintenance:** Digital machines, like gas machines, break. It is a reality of complex systems. This fact places the burden on preventing or minimizing outages because of maintenance and repairs. Many of the digital machines and devices will be designed to predict and prevent failures, but ultimately, this must become a core competence of the farmer or his IT back office. Given that each farm will use the machines differently, the maintenance needs will likely be unique.

- **Predictable productivity:** In today's farms, the yield and productivity of crops vary significantly. Whether it is the weather, impacts of deforestation, or the impact of certain pesticides and fertilizers, it is an often-unpredictable environment. By 2020, productivity will be more predictable. Given all of the sources of data, GIS and GPS capabilities, and the intensive learning that will happen over the next five years, yields will become predictable, creating greater flexibility in the financial model for a farmer.

- **Risk management:** In 2020, instead of being a key determinant of success, weather will simply be another variable in the overall risk management profile of an asset (in this case, a farm). Given the predictable productivity, risk management will be more about managing catastrophic outliers and, in some cases, sharing that risk with counterparties. For example, index-based insurance offers great potential in this area.

- **Real-time decision-making:** Decisions will be made in the moment. With the growth of streaming data, farms will be analyzed as variables are collected and acted upon immediately. Issues will be remediated in 2020 faster than they will be identified in 2014. This is part of what drives the predictable productivity.

- **Production variability:** Farms will no longer produce a single crop or focus on a subset. Instead, they will produce whatever will drive the greatest yield and productivity based on their pre-planting-season analysis. Farms will also begin to factor in external data sources (supply and demand) and optimize their asset for the products in greatest demand. This will completely change the variability that we see in commodities and food prices today.

CALIFORNIA, 2013 (CONTINUED)

As I left the headquarters of the agricultural company outside of San Francisco, I was amazed that a belief persists, in some places, that weather is *the* major force impacting our ability to grow consistent and productive crops. That does not seem much different from the pioneer farms of the 1800s, where the weather determined not only their business, but also their livelihoods.

Perhaps, as postulated before, that is the easy answer, as opposed to the real answer. The innovations that we've seen with precision farming, using data to transform potato crops, and the emergence of leaders like Monsanto, makes it evident that the weather is merely one variable that could impact crops in the future.

Data trumps weather. Farming and agriculture will be transformed by making the leap of acknowledging this truth.

2

WHY DOCTORS WILL
HAVE MATH DEGREES

UNITED STATES, 2014

WE ALL KNOW how it works. You walk into a doctor's office complaining about some pain in your leg or otherwise. They take your temperature, get you on the scale, check your blood pressure, and perhaps even get out the rubber hammer. These measurements are simply snapshots at one particular instant in time and may be subject to error. This limited dataset fails to capture temporal variations or the many other important factors that are required to assess the patient's health status. After reviewing the few measurements collected, the consultation between the patient and doctor begins. There is typically a discussion of symptoms or problems, with the physician utilizing her judgment and experience to assess the situation. While we have more advanced imaging technology and surgical equipment, these tools are rarely used early in the consultation process. Accordingly, based on the rudimentary physical analysis, along with the discussion with the patient, the physician will assert the condition that they believe is present, followed by a recommended treatment.

This approach, which is common throughout the world, is much more based on instinct and gut feeling than a scientific approach to analyzing data. Accordingly, it seems that most decisions are made based on the opinion of the physician instead of a data-proven truth. This type of opinion-based medicine is a problem in both doctor-patient care and in medical research. This is a symptom of a lack of data, as well as years of training physicians to perform without complete data.

The data collected in a typical office visit is only a fraction of the data that could be collected if health were viewed as a data problem. And, if health was redefined as a data problem, physicians would likely need different skills to process and analyze the data.

THE HISTORY OF MEDICAL EDUCATION

Universities first began cultivating physicians in 1220 in Italy. The first medical school was Schola Medica Salernitana in Italy. The students, primarily from wealthy families, often studied for eight years to earn a degree and the designation of doctor. Universities, and medical education, continued to expand around Europe, with schools being founded in France, England, and other parts of Italy. Regardless of location, curricula focused on the writings of Hippocrates, Galen, and Aristotle, to name a few. Accordingly, the course was heavy on theory and reflection, while light on clinical work and analysis. The students were trained in theory, not fact. That set the tone for the next 800 years.

SCIENTIFIC METHOD

The 19th century is the first time that the scientific method was applied to medical research, leading to a broad set of advancements. The scientific method is about seeking knowledge (or data) to either disprove or prove a theory. This is important in the annals of medicine, as it was a substantial move beyond theory, into fact-based practice.

The scientific method involves formulating hypotheses and systematically researching, reviewing, and challenging those hypotheses. The steps that follow are not sequential, and perhaps not each one is needed. They are principles to guide thinking. The steps are:

- **Formulation of a question:** Posing a previously unanswerable proposition.
- **Hypothesis:** A statement of the answer to the question, without worrying about whether the statement is accurate or not.

- **Prediction:** A contemplative view of what the data will prove the answer to be.
- **Testing:** Extensive trials to assess variables and factors related to the hypothesis.
- **Analysis:** Compilation and reconciliation of the data derived from testing.

The entire field of science is based upon this methodical approach to analysis. Yet, it seems modern medical practices have drifted away from this rigor. Due to time limitations, the physician is expected to use qualitative information, based on a visual inspection of the patient and a series of questions, to replace a more thorough data collection exercise. Recent advances in technology for collecting big data and an ability to process data in real-time has the potential to address this shortcoming.

RISE OF SPECIALISTS

The American Medical Association was not established until 1847, which was over 600 years since that first school in Italy. It was 1933 when examining boards for the certification of specialists was approved.

In 1910, Abraham Flexner first published a report on the state of medical education in the U.S. and Canada, which concluded that the then-current institutions were substandard. His three chief suggestions in this report were that the schools should have minimum admission standards, should be four years in duration, and "proprietary" institutions should be dissolved or combined into universities. On the heels of this report, 89 institutions were closed, and the number of physicians fell from 173 per 100,000, to 125 per 100,000. The impact of Flexner was not trivial, as it created a scarcity of physicians.

In 1959, the Bane Report found that the U.S. was facing a physician shortage of approximately 40,000. As a result of this finding and a new focus on education, the number of physician graduates rose from 7,849 in 1965 to 16,935 in 1980, where it has remained essentially unchanged. The quick rise in physicians resulted in the rise of specialists, as many physicians had to specialize in order to differentiate their skills.

A specialty in medicine is a component of medical science, typically just with a narrow focus. Many physicians who complete medical school today will often extend their education into a specific field of interest. Specialties are often designated in the following categories:

- **Surgical Medicine:** Focused on invasive surgery versus less invasive diagnosis and treatment. This will typically include a specialty in a certain condition.
- **Internal Medicine:** Focused on diagnosing and treating adults, including disease prevention.
- **Age:** Specializing in a certain age range of patients, sometimes with a specialty in certain type of condition.
- **Diagnostics or Therapeutic:** Specializing in therapy and the long-term management of a condition (versus focused on assessing/diagnosing a condition).
- **Organ-Based or Technique-Based:** Specializing in a certain organ of the body and/or a certain technique for treating an organ (or organs).

Specialties are necessary, as we have learned more about many conditions and understand that different sets of knowledge and treatment are required, depending on the condition. However, the emergence of specialties leads to a set of significant implications:

- Medicine has become less efficient over time, as you often have to visit multiple specialists to receive the right diagnosis.
- Specialties, by definition, create biases. A doctor who has spent 20 years working on hips is more likely to see or diagnose a hip injury, even if symptoms indicate other possibilities. It's human nature.
- Specialists are frequently paid more, leading to more specialists and, in turn, more biases.

With the increase in the number of physicians, and corresponding rise in specialists, modern medicine is being governed by human judgment (opinion and bias), instead of data-based science. Nearly $1.2 trillion of U.S. Gross Domestic Product (GDP) goes to healthcare, while approximately $500 billion of that is spent on doctor's visits, 25 percent of which are just informational in nature. There are too many opinions and not enough data science in medicine today.

WE HAVE A PROBLEM

Specialties are necessary, as we have learned more about the variety of conditions that exist and understand that different sets of knowledge and treatment are required, depending on the condition. However, the emergence of specialties leads to a set of significant implications, as the following sections illustrate.

BEN GOLDACRE

Ben Goldacre was trained in medicine in Oxford and London. He is 40 years old, younger than the average doctor or professor. He is also a hybrid, working as an academic in epidemiology. Ben has created waves in the field of medicine, as he systematically breaks down the use of science and statistics in medicine. His first book, *Bad Science* (Harper Perennial, 2008), has sold over 500,000 copies.

His 2012 TED talk, "What doctors don't know about the drugs they prescribe," is perhaps most impactful when making the case that there is a widespread selection bias in academic publishing. Or, as Goldacre calls it, *publication bias*. He cites two examples, the first of which illustrates basic selection bias, while the second demonstrates publication bias.

- **Nostradamus:** Many people believe he predicted the future, but they ignore the hundreds of other people that predicted things that did not happen. They pay attention only to the person who was right (perhaps out of luck), as opposed to all those who were wrong.
- **Medical trials:** A publication will cite a study showing that a drug cured a certain medical condition in eight out of ten people tested. However, that publication conveniently ignores ten other tests, of the same drug and same condition, where only two out of ten people were cured.

Publications, or media more broadly, will often focus only on the things that people got right, instead of all the things that they got wrong. And, as Goldacre asserts, we will continue to hear only about positive outliers until it becomes easier to publish negative results in science and medicine.

Publication bias is the notion that unflattering data gets lost and is never published. Goldacre has found it to be pervasive in all fields of medicine and that approximately half of all trials are buried. Even worse, he suggests that positive findings are twice as likely to be published as negative findings. Simply put, nobody wants to hear bad news.

Bias and opinion continue to encroach on the modern medical field. In the Data era, a decision would never be made while only comprehending a subset of the data available. Yet that is exactly what Goldacre has uncovered in the medical field today.

VINOD KHOSLA

Vinod Khosla is one of the most successful venture capitalists in the history of Silicon Valley. He was an original founder of Sun Microsystems, and has since gone on to finance a variety of start-up companies as a venture capitalist. While he is not a medical expert, he is a data expert. In his speech at Stanford Medicine X, Khosla highlights three major issues in medicine today:

- **Doctors are human:** Doctors, like everyone else, have cognitive limitations. Some are naturally smarter than others or have deeper knowledge about a particular topic. The latter leads to biases in how they think, act, and prescribe. Most shockingly, Khosla cites that doctors often decide on a patient diagnosis in the first 30 seconds of the observation. Said another way, they base their diagnosis on a gut reaction to the symptoms that they can see or are described to them.

- **Opinions dominate medicine:** Khosla asserts that medicine is much more based on opinion than data. He cites the Cleveland Clinic Doctors' Review of Initial Diagnosis study, asserting that Cleveland Clinic doctors disagree with initial diagnoses 11 percent of the time. In 22 percent of cases, minor changes to treatment are recommended. And in a startling 18 percent of cases, major changes to treatment are recommended. As Khosla states, "This means it's not medical science."

- **Disagreement is common among physicians:** As alluded to in the preceding bullet, doctors disagree a lot. It's so dramatic, that, Khosla states, "whether or not you have surgery is a function of whom you ask."

Medicine is currently a process of trial and error, coupled with professional opinion.

THE DATA ERA

The Data era in medicine will be defined by a shift from intuition and opinon to data. We can collect more data in a day now than we could in a year not too long ago. Collecting data and applying it to solve healthcare problems will transform the cost and effectiveness of medicine. The question is how quickly we can get there.

COLLECTING DATA

At this stage, we are just starting to experiment with the impact that data can have on medicine. There are obvious use cases, such as medical imaging, where we know that improved data and insight can lead to better answers. But there are also less obvious applications of data to medicine, if we do not confine ourselves to data that is relatively easy to access and utilize.

Peter Diamandis

Peter Diamandis was born in the Bronx, New York in 1961. His Greek immigrant parents were both in medicine. Peter was interested in a broader frontier: space. He began competing in local competitions to build rockets, incorporating unique or innovative designs. This passion eventually led him to MIT, and then on to Harvard for medical school. While at Harvard, he founded the Space Generation Foundation and eventually founded the X PRIZE Foundation, which awards inspiring entrepreneurs innovating in space flight. With his background in biology, physics, and medicine, coupled with his passion for space, Peter is very much a data guy.

In 2012, the X PRIZE Foundation and Nokia co-launched a challenge, The Nokia Sensing X Challenge, to award the person who can develop the most innovative biosensor. This challenge will take place over three years in a series of three events, with the winner being chosen in 2015. Biosensors collect, monitor, measure, and act on health data generated by a human body. This contest alone could propel the industry forward in terms of innovation.

In Diamandis' book *Abundance: The Future is Better Than You Think* (Free Press, 2012), he talks about how long it takes to train doctors — and even worse, once they are trained, much of what they have learned may be obsolete. Diamandis envisions a future where "you will have the data and data analysis to become the CEO of your own health." Biosensors, as envisioned in the 2012 challenge, enable this transformation.

The ability to collect, monitor, and measure thousands, if not millions, of data points generated by a human being changes the platform for medicine. It drives the shift from opinion to data and from theory to fact. It reverses the approach that was first put into motion in the 1200s at Schola Medica Salernitana in Italy.

CellScope

Medical data can be harvested nearly anywhere. With modern tools and technologies, it often just requires the initiative of an individual. CellScope is a company founded in 2010 whose first product offering will be a smartphone-enabled otoscope.

The otoscope is a medical device that can diagnose ear infections. Once a child is in the doctor's office, the doctor can quickly assess an ear condition and decide on the appropriate treatment, using the otoscope. The common nature of ear infections leads to nearly 30 million doctor visits per year in the U.S. alone. But what if a parent could collect this data in the home and share it with a physician? Giving individuals this capability would change medical practice as we know it in that it could be done remotely and conveniently. With CellScope, parents will attach a peripheral to a smartphone and send an image of the child's inner ear to a physician. The physician can remotely and quickly make an assessment of the condition and write a prescription if necessary.

Since the physician will not actually see the child, their diagnosis will only be based on data, eliminating any biases or opinions that may come from a consultation.

Simply having the ability to collect more data will change medicine. But data collected is not the same thing as data utilized. Accordingly, new applications will emerge that will enable innovation with the new datasets.

TELEMEDICINE

A new field of research and innovation has arisen from the availability of sensors, mobile telephones, and social media to collect a wide range of data. *Telemedicine* refers to the ability to provide healthcare at a distance using information and communication technologies. While having the potential to generate big data, this new branch of medicine also promises to enable and promote self-management of a range of medical disorders, including neurological disorders, diabetes, asthma, and cancer.

True Colours

Collaboration between psychiatrists, engineers, mathematicians, and data scientists at Oxford University is helping to revolutionize the way medical information is collected and analyzed. Working with the U.K.'s National Health Service, True Colours (`www.truecolours.nhs.org.uk`) offers individuals an online self-management system that allows them to monitor their symptoms and experiences using text, email, and the Internet. The system is designed to identify even small changes in health and wellbeing for a wide range of different conditions, from post-operative quality of life to long-term mood disorders. By answering questionnaires, a person can create a record of how they are feeling and can see how it changes over time. This system helps individuals to manage their own health and to share information with their family, friends, or care team.

A testimonial (`https://truecolours.nhs.uk/www/info.jsp`) from Professor John Geddes, a psychiatrist and clinical practitioner involved in creating True Colours, demonstrates just how much impact the system is having:

From a clinical point of view the effect of the system has been dramatic. We feel much better informed about how well our patients are and about estimating the effects of changes in treatments.

Similarly, patients have been extremely positive about the difference that True Colours has made to their lives. One patient (`https://truecolours.nhs.uk/www/info.jsp`) made this very clear:

In a whole host of ways this has proved to be a godsend. The regularity with which the prompt text arrives - infallibly each Monday, before noon - acts as a support, and indeed comfort, at times when illness seems to bring only irregularity and uncertainty.

Quantitative analysis of the mood data collected by True Colours from large cohorts of patients is already helping to characterize the nature of mood disorder more accurately. The data provides a means of classifying patients based on the evolution of mood ratings over time. Combining the mood data with external factors such as the environment and social media is helping to better understand and stabilize mood variability. This mood data also has potential to evaluate different therapies and facilitates clinical trials for assessing the viability of drugs for treating mood disorders.

The technology already exists to collect the medical data and to help physicians to arrive at a final diagnosis using prior knowledge and data analysis. Access to data is only the start as the development of enhanced data analytics is also likely to improve the medical field. T. Tsanas, M. Little and P.E. McSharry outline a methodology for the analysis of medical data (*Handbook of Systems and Complexity in Health*, Springer, 2013). By applying sophisticated machine learning techniques as opposed to traditional regression analysis, they show that is possible to reduce misclassification rates by as much as 36 percent on average across six existing medical datasets.

Telemedicine systems such as True Colours offer great opportunities for physicians to use data to obtain a preliminary assessment of their health even before they enter the doctor's surgery. It is not a great stretch of the imagination to see how this approach to data collection could be used to create an effective tailor-made early warning system for individuals.

INNOVATING WITH DATA

Utilizing data to drive innovation in the medical field is largely reliant on technology transfer. Much of the technology needed for major innovations in medicine already exists; it just needs to be repurposed and applied. Over time, new technology will be needed, developed, and exploited. But for the moment, the largest opportunity lies in technology transfer.

GRoK

NASA has a technology transfer program which unleashes NASA's existing investments in research and technology to the public. Under the agreements from NASA, companies can gain access to certain technologies and patents, which can be used in the development of new commercial applications. In February 2014, this is exactly what GRoK Technologies, LLC did.

As a licensee of NASA, GRoK is seeking to improve human lives through a novel application of technology and data: regenerating bone and muscle. GroK is pursuing two platforms, harnessing the NASA innovations:

- **BioReplicates:** Users can create models of human tissue, which can be used for a variety of purposes, including testing cosmetics, drugs, and anything that could be perceived to have side effects on human tissue. The data collected from this platform will be significant as it is tested over time.
- **Scionic:** This platform is targeting, treating, and minimizing pain and inflammation in humans. The approach would be non-invasive and non-pharmacological. Instead, it is based on collecting data to assess a person's condition.

Quanttus

Cardiovascular disease causes more deaths per year than any other disease. Cardiovascular disease kills twice as many people as cancer and over three times the amount of deaths due to age-related illnesses. Cardiovascular disease is human enemy number one and is so tragic for two reasons:

- Cardiovascular disease often develops slowly and methodically over time. Because of this, there are rarely indications that it is lurking.
- Cardiovascular disease will sometimes strike suddenly (in the form of a heart attack), and the victim is dead before he or she is even aware there was a problem.

For these reasons, fighting cardiovascular disease is really about identifying it and managing it over time. Neither of these approaches are common in traditional medicine, and accordingly, they call for a different approach.

Inspired by XPRIZE, Shahid Azim founded Quanttus with the mission of utilizing data to manage the life cycle of cardiovascular disease. An early prototype of the company used a variety of wearable sensors to measure vital signs ranging from respiration to heart rate to blood pressure. According to Vinod Khosla, an investor in Quanttus, "The team's approach to using real-time physiological data will create powerful new tools for consumers and healthcare providers that will change how we understand our health."

The body has the ability to produce millions of data points in real-time. We are only limited by our ability to collect, process, analyze, and act on that data in real time. The skills required by doctors will change as data is more widely collected. Physicians will have to possess the skills to derive insights from the data.

HealthTap

Even if a person were to collect millions of data points per day from his body via a Quanttus-like monitoring tool, the insight would still be limited to that single dataset, from that single person. This insight reveals that there is a social aspect to medicine — analyzing datasets over a large population could perhaps lead to even better outcomes.

HealthTap is building a patient-focused health information service, starting with a mobile application that anyone can run on most any device. For a patient, the experience is simple: Tap a button, ask a question, get an answer. Behind the scenes, HealthTap is processing vast datasets, using machine-learning, natural language processing, and voice recognition, to find answers to the questions that are asked. It's almost like having a doctor in your pocket, anytime you need her. There are currently over 55,000 doctors using the application and providing answers. Like any network, as the users (patients and doctors) increase, the value of the network increases, often exponentially.

Data is utilized in many forms on HealthTap. The two most prominent are in the routing and answering of questions:

- As questions come in, the question is analyzed by an algorithm and routed to the physician that can best answer that question.
- Previous questions and answers are analyzed to provide self-service to patients, based on certain symptoms or conditions.

IMPLICATIONS OF A DATA-DRIVEN MEDICAL WORLD

A medical industry that starts with the collection of data and ends with science-based analytics of that data is very different from how the industry began many centuries ago. This fundamental transformation will impact every aspect of healthcare:

- **Biotech:** Biotech is the one corner of the healthcare industry that has a head start on the usage and application of data. The primary opportunity for improvement comes from greater access to data as the other components of the value chain mature in their collection and use of data and share the data broadly.

- **Pharma:** Pharma today is largely trial and error, based on hypotheses. This approach has led to some tremendous discoveries over the years, but is perhaps not as efficient as it could be. If the data available to pharma for clinical trials or even prior to clinical trials was 100 to 1,000 times as great as it is today, success and time to success would improve.

- **Payors:** Payors are in the business of insuring groups or individuals and then deciding how much of that cost to share with the patient. Data promises to improve the efficiency of the payors' role in the lifecycle of a patient.

- **Providers:** Any time a person walks into a new physician's office, he is a stranger. He becomes a known quantity through what he tells the physician, but he is rarely sharing facts; he is sharing perception. Contrast this state with a world where his medical history, in data, is provided the minute he walks into the office. His chances of receiving the proper care increase exponentially.

- **Patients:** Healthcare can transform for patients when all providers have access to pertinent information and can correlate data across different visits and conditions. When medicine becomes personalized, it becomes more effective.

THE FUTURE OF MEDICAL SCHOOL

Medical schools must evolve as technology advances. Most advancement in medical schools, based on technology, have been focused on utilizing advanced tools and equipment, as opposed to addressing the core knowledge needed by a physician in the data era.

A TYPICAL MEDICAL SCHOOL

Most medical schools are a four-year program. A student spends the first two years in classrooms and labs, preparing for the first set of licensing exams, known as the United States Medical Licensing Exam (USMLE). Once they pass those, they proceed to years three and four, where clinical rotations, coupled with specialty courses, is common.

The curriculum for the first two years of medical school varies by school, but typically includes topics such as:

- Anatomy
- Biochemistry
- Genetics
- Microbiology
- Pathology
- Pharmacology

The curriculum is heavy on the sciences, the human body, and the human condition. This has been typical since the first medical schools in the 1200s.

At the end of years three and four, there is a second licensing exam, which extends to assess the clinical and diagnostic skill of a student. The last test typically comes during the students' residency, and with the successful completion of this third part of the USMLE, the student becomes licensed as a physician.

All this time and investment, yet the newly minted physician is unprepared for practicing in the data era.

A MEDICAL SCHOOL FOR THE DATA ERA

The basic physician skills will still be both taught and necessary for physicians. Whether it's anatomy, microbiology, or genetics, this base of knowledge will never be replaced by technology. But the data era requires an augmentation in curriculum to include key skills required for data-based analysis:

- Mathematics
- Statistics
- Probability
- Data Analysis and Tools

Each of these are examples of the base of knowledge that will be required for the future. As described earlier in this chapter, with HealthTap, the physician only receives data and with that data, they are expected to offer a prescription. That is very difficult to do, if a physician has only been schooled in hands-on clinical work.

In the case of Quanttus, the monitoring equipment produces streams of data. As the capabilities of devices are enhanced, it's expected that the data will increase from millions of data points to tens of millions of data points. Analyzing large data sets, to prescribe a treatment, will demand more than simple knowledge of pharmacology.

The skills of physicians will necessarily evolve in the data era, and that has to begin in medical schools. This focus will expedite the move away from opinion-based medicine to a future that the ill prefer: prescriptions based on hardened data analysis.

UNITED STATES, 2030

We all look forward to how things will likely be in 15 years. You walk into a doctor's office, and the physician immediately knows why you are there. In fact, she had discussed some data irregularities that she had spotted at your annual physical exam, six months prior. At that time, she told you that there was a 20 percent chance it was nothing, and 30 percent chance that it was related to your change in diet, and a 50 percent chance that your low blood sugar was becoming problematic.

She doesn't need to take your temperature, as she receives that data direct from your home every day. You also take your own blood pressure monthly and that is transmitted directly to your physician. Instead, the discussion immediately turns to the possible treatments, along with the probability of success with each one. Recent data from other patients with a similar history and physiology indicate that regular medication will solve the issue 95 percent of the time. With this quick diagnosis, involving no opinions, you are on your way after ten minutes, confident that the problem has been solved. This is medicine in the data era, administered by a physician steeped in mathematics and statistics.

3

REVOLUTIONIZING INSURANCE: WHY ACTUARIES WILL BECOME DATA SCIENTISTS

MIDDLE OF SOMEWHERE, 2012

INSURANCE COMPANIES ARE unique. It seems that a large portion of them are headquartered in fairly remote regions of the world, away from the dynamism of New York, Chicago, Los Angeles, or other large cities. Are they a product of their location — conservative and measured? Or is their location a product of them? After all, most insurance companies in fairly remote areas are the dominant employer in the area. Culture is important and location says a lot about culture.

Most insurance companies have been around for a long time. The Hartford was founded in 1810, MetLife was founded in 1868, Prudential was founded in 1875, State Farm Insurance was founded in 1922, and Geico started operations in 1935. Given the age and maturity of these companies, they possess many commonalities:

- **Application sprawl:** They have hundreds of applications and tools for managing the insurance life cycle, ranging from underwriting systems, to policy administration, to customer relationship management.
- **Diverse policies:** They offer many types of insurance: Auto, home, fire, flood, and many others. In some cases, they have thousands or millions of each type of policy.
- **Multiple channels:** A mix of online, agency, indirect, and direct contact with consumers.

Perhaps the word that captures the commonality in long established insurance companies best is complexity. The insurance world displays complexity in form, diversity, tenure, and adaptability. After all, insurance companies have survived throughout the years by executing a particular recipe and rarely, if ever, deviating.

SHORT HISTORY OF PROPERTY & CASUALTY INSURANCE AND UNDERWRITING

Insurance practices date back to 1700 BC, with continued innovations across the centuries since then. There is evidence of this rich history in modern day insurance today. Here is a timeline of significant occurrences in the history of insurance:

- **1700s BC:** One of the first documented applications of insurance dates back to 1760 BC in Babylonia. The Babylonians established an early trade culture in what is now modern day Iraq and developed one of the earliest forms of long form writing with the Babylonian Code of Hammurabi. At the time, if a person were to ship goods for trade, he would frequently have to borrow money in order to charter the ship and transport the goods in the hope of securing their safe arrival. For an additional fee, that same person could ensure that he would not have to repay the loan if the ship was lost or sank. In a sense, it was marine insurance, before such a term existed.
- **1300s:** The very first insurance policy was signed in Genoa, Italy in 1343. It was similar to the practice in Babylonia, except it involved signed paper and hence, a contractual obligation of parties.
- **1600s:** The practice and laws of probability came into being at this time. Thanks to the mathematical work of Blaise Pascal and Pierre de Fermat on probability, early insurance companies began to focus on risk management and pricing optimization. Insurance came into sharper focus after the London fire of 1666, which resulted in the birth of the first fire insurance company.

- **1700s:** Benjamin Franklin founded the Philadelphia Contributorship, leading the charge of the creation of fire insurance in the colonies, in 1752. Franklin is sometimes referred to as the father of insurance, as eventually he extended beyond fire insurance to offer insurance for crops, life insurance, and insurance for widows and orphans. In its first year of operation, the Contributorship wrote 143 policies.
- **1800s:** One of the first car insurance policies was written in 1897 in Massachusetts. This became a catalyst to the broader application of insurance to manage one's financial security and remove uncertainty.

The history provides context to show how modern day insurance has evolved into its current form. Through time, insurance has always been about sharing and managing risk (for a price). Risk comes in many forms. Can data change how we think about insurance and risk?

ACTUARIAL SCIENCE IN INSURANCE

Insurance today takes many forms: life insurance, health insurance, property insurance, casualty insurance (liability insurance for negligent acts), marine insurance, and catastrophe insurance (covering perils such as earthquakes, floods, windstorms, and terrorism). You name it, you can probably insure it. A recent example of something you might not think of as insurable is the $1 billon prize promised to any individual that could correctly pick every winner in the NCAA Men's Basketball Tournament. While QuickenLoans sponsored the contest, the insurance underwriter behind it was Warren Buffett's Berkshire Hathaway.

Prior to the insurance innovation that led to the varieties just mentioned, a branch of management science was established that served as the root of insurance: actuarial science. Actuaries are in the business of assessing risk and uncertainty. Said a different way, they value and assess financial impact of a variety of risks. But that is much easier said than done. A variety of inputs provide the information an actuary needs for better decision-making.

In 1762, the Society for Equitable Assurances on Lives and Survivorships was founded as a mutual company. This company created the impetus for the actuarial profession. Through the years, many organizations have emerged, dedicated to the practice.

The Institute and Faculty of Actuaries (IFoA) is the only professional organization in the United Kingdom dedicated to educating, regulating, and generally advocating for actuaries worldwide. If Ben Franklin is the father of insurance, Chris Lewin, associated with IFoA, is probably the original actuarial historian. He has published regularly on the topic and is well known in the community for his significant contributions. Lewin offers a compelling point of view in a 2007 seminar presentation titled, "An Overview of Actuarial History" (http://www.actuaries.org.uk/research-and-resources/documents/overview-actuarial-history-slides-notes). He explores how the craft developed over the past centuries and concludes that while data has always played in a role in insurance in history, we are likely at the dawn of a new data era for the industry.

In his 2007 presentation, Lewin defines the role of the profession:

> *Actuaries mathematically evaluate the probability of events and quantify the contingent outcomes in order to minimize the impacts of financial losses associated with uncertain undesirable events.*

Even more simply, Lewin commented, "An actuary looks at historical data, and then makes appropriate adjustments (subjective, of course)." One of the primary skills of an actuary, therefore, is to make estimates based on the best information available. In the case of life insurance, actuaries have to estimate a variety of factors:

- **Cash flows:** Expected future cash flows, from a dollar invested today, assuming a variety of potential factors.
- **Demographics:** Estimate risks based on life expectancy, which includes factors such as sex, age, tobacco-usage, and overall health condition.
- **Profession:** Estimate risks based on the insured's chosen line of work, including the odds that the insured will not change professions.
- **Probability:** Estimate the probability of a life-changing event, given the insured's demographics, profession, and family history.

Even if an actuary uses data to develop an informed judgment, that type of estimate does not seem sufficient in today's era of big data. There is something about modern-day insurance that has led the industry to believe that informed judgment is good enough. As the quantity and quality of data improves, it will be possible to calculate increasingly accurate estimates based directly on information, negating the need for human judgment and associated biases.

Lewin believes that the four components shown in the following figure contributed to the creation of actuarial science.

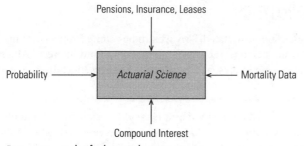

Four components that feed actuarial science

Lewin's insight does not stem merely from his own understanding of the
components of the theory. In fact, he credits many thinkers throughout the
years for their contributions in each area:

- **Compound Interest:** Richard Witt
- **Probability and Probability Theory:** Thomas Gataker and Christian
  Huygens
- **Mortality:** John Graunt, John Hudde, Jan de Witt, and Sir Richard
  Corbet
- **Demographics:** Sir William Petty
- **Mortality Data:** Caspar Neumann
- **Mortality Analysis:** Edmond Halley

By reviewing the work of all of these seemingly unconnected experts,
spanning centuries, Lewin highlights the lineage of actuarial science. The
following sections offer a summary of that research.

PENSIONS, INSURANCE, LEASES

A pension is a fixed amount paid to a person, on a periodic basis, typically
after he retires or reaches a retirement age. The concept of pensions dates
back to 582 BC, but it wasn't until 1875 in the United States that American
Express established the first private pension plan. Insurers entered into the
pension arena in 1921, when the Metropolitan Life Insurance Company
issued its first group annuity contract. Pensions, insurance, and leases all
became inputs into actuarial science since they represent future payment
obligations, which have to be considered in assessing risk.

COMPOUND INTEREST

Warren Buffet once famously stated, "My wealth has come from a combination of living in America, some lucky genes, and compound interest." Albert Einstein cited compound interest as "the greatest mathematical discovery of all time."

Compound interest is best understood as it is taught in elementary schools today, with a simple question: "Would you rather (a) receive $10,000 per day for 30 days or (b) a penny that doubles in value for 30 days?" Most children choose option (a), until they realize that option (b) results in $5 million dollars in 30 days, while option (a) results in a measly $300,000. This illustrates the power of compound interest. This is why some people call compound interest the eighth wonder of the world.

Richard Witt, a London mathematical practitioner, published the first English book on compound interest in 1613. He shared many examples of the concept, including the mathematical basis behind it. This became a fundamental concept, feeding into actuarial science, as he explored the impact of payments and returns.

PROBABILITY

Probability theory dates back to the 16th century. Gerolamo Cardan first wrote about the topic as related to dice games, but it was many more years before there was a wider-spread publication of the topic. Probability directly feeds into actuarial science, as it addresses the chance of any risk occurring. The chance of something happening, moderated by the expected outcome, is the essence of an actuary's work.

MORTALITY DATA

In the 1600s, large cities typically published a daily Bill of Mortality, which served as a caution of diseases or potential contagious conditions. Over time, as more and more cities began to publish such bills, cities could ascertain patterns and use that data to help prevent certain diseases and to forecast potential plagues. According to Lewin, John Graunt's life table of 1662 drew "attention to the regularity of the patterns of life and death." With this knowledge, he could predict the population sizes of the time and how that would fluctuate seasonally or over different time periods.

Lewin's insight into the relationship of these components was prescient, in terms of building a model that feeds into our modern day actuarial science. With the understanding of patterns of life and death (mortality), the likelihood of certain events (probability), an understanding of cash flows (pensions, leases, and insurance), and the ability to estimate the growth of cash flows (compound interest), the building blocks of actuarial science fall into place.

MODERN-DAY INSURANCE

Modern-day insurance has slowly begun to accept new approaches to managing risk and optimizing customer relationships. The innovation in insurance over the last 15 years has been significant, with the automation, acceleration, and in some cases elimination, of key business processes.

EIGHT WEEKS TO EIGHT DAYS

Back in 2002, a major insurance company was exploring how to transform their claims underwriting process from eight weeks to eight days. This company, recognizing the data available to them and understanding client preference, moved aggressively to change their processes for improved speed and accuracy. The critical insight was that analytics had to start with process; if they didn't understand why their current process took eight weeks and how the inputs to that process affect the timeframe, they could not transform to eight days. For this company, the process findings were substantial:

- **Workflow:** The main driver of the eight-week cycle was workflow and approvals. A clerk would enter claims data and manually pass the form along to the next person. In most cases, there were 15 to 20 checkpoints along this process. So, even if everything was perfect, it would take a minimum of six weeks. But, as you can imagine, with all the manual inputs, data was frequently missing or more information was required. Every time data was needed, the process would restart at step one. With the current process, eight weeks was actually a miracle, as it seemed that it should take much longer than that.
- **Data collection and entry:** Data collection and entry was all manual. If a client filled out a form, that form was rekeyed into a specific system to start the process. Data was not ingested at each step; it was recreated at each step. This led to errors and loss of productivity, which were both key contributors to the entire process requiring eight weeks.
- **Decision making:** Decisions were ultimately based on experience and informed judgment, using the limited data available. This is quite different from making decisions based on data, and doing so empirically.

Recall that this was 2002. There was not enough data available nor a data culture to attack the third item.

ONLINE POLICIES

Much has changed since 2002. Since then, online presence has become a primary route to market for insurance (at least, consumer insurance). This enables consumers to receive insurance quotes online, within minutes, which is in stark contrast to the pre-Internet process, which often required multiple face-to-face meetings before a person could secure insurance. The ease with which policies can be quoted and obtained online increased competition, which drives rates down and, accordingly, creates more risk for an insurance company. In other words, there is a risk of policy stacking, whereby an insurance company writes multiple policies for the same person, leading to a large exposure, without the ability to spread that risk across a number of individuals.

The openness of the Internet means that an insurance company could write, in less than a day, a set of policies that would create a sizeable risk exposure. While this is unlikely, it is inherently possible. Hence, this is the other side of the fast-and-convenient coin: risk.

Modern-day insurance has seen tremendous advances in business processes, decision-making, and time to market. With each of those advances, there is a corresponding risk that must be managed. Only as we enter the Big Data era can both be managed for the optimal outcome.

THE DATA ERA

The data era has already begun to spark a new wave of innovation in insurance. As we saw in farming in Chapter 1, "Transforming Farms with Data." rich access to a variety of data assets, coupled with the ability to analyze and act, enables processes that were not previously possible. This will usher in the era of dynamic risk management and improved approaches for modeling catastrophe risk.

DYNAMIC RISK MANAGEMENT

In the case of automobile insurance, the industry commonly refers to this type of insurance as Usage Based Insurance. There are two types of policies under this type of insurance — Pay-As-You-Drive (PAYD) and Pay How You Drive (PHYD). However, dynamic risk management can apply well beyond the scope of driving and automobile insurance.

Dynamic risk management is an accelerated form of actuarial science. Recall that actuarial science is about collecting all pertinent data, using models and expertise to factor risk, and then making a decision. Dynamic risk management entails real-time decision-making based on a stream of data. Let's explore the two models with an example of car insurance for a 22-year-old female:

- **Actuarial insurance:** Collect all the data available for the 22 year old — her driving history, vehicle type, location, criminal history, etc. Merge that data with demographic data for her age, gender, location, and work status. Leverage methods like probability, mortality, and compound interest to estimate benefits and obligations. Then, offer a policy to the woman, based on these factors.
- **Dynamic risk management:** Install a sensor in her car and tell her to go about her normal life. Collect mileage, time of day she drives, how far she drives, acceleration/deceleration, and the locations that she drives to. When she is driving, monitor the motion of the vehicle. Said another way, this is an on-board monitor, constantly pricing her insurance policy based on her personal driving behavior. If she drives well, her next premium may be lower. The policy is tailor-made for her and is based on actual data, as opposed to estimates.

There is now increased momentum for dynamic risk-management. In March 2011, the European Court of Justice stated that "taking the gender of the insured individual into account as a risk factor in insurance contracts constitutes discrimination." Since December 2012, insurers operating in Europe are no longer able to charge different premiums on the basis of an insured person's gender. There are good reasons why insurers might want to use gender as a means of quantifying risk. Men under the age of 30 are almost twice as likely to be involved in a car accident as their female counterparts. Insurers also have empirical evidence to show that the claims they receive for young men are over three times as large as those for women.

Arguments around gender equality have rightly determined that it is unfair to blindly discriminate against young men. Furthermore this debate has highlighted the need for more appropriate metrics for forecasting risk rather than the blunt use of gender. This gap in the market calls for better models and dynamic risk management based on the actual driving ability of the individual. Unfortunately, in the meantime, we are all paying the price as car insurers have increased their premiums across the board.

Currently, many of the large insurance carriers offer some version of dynamic risk management, or pay-as-you-drive insurance for automobiles: Progressive, Allstate, State Farm, Travelers, Esurance, the Hartford, Safeco, and GMAC, to name a few. Most of the insurers market that premiums will cost 20 to 50 percent less for consumers who adopt this approach. The National Association of Insurance Commissioners estimates that 20 percent of insurance plans will have a dynamic approach, by 2018. For the moment, dynamic insurance for automobiles is less than one percent of the market.

Dynamic risk management can apply to any data-centric insurance process, whether a company is leveraging telematics or data points about a consumer in a lending scenario. In the Big Data era, dynamic risk management will become routine.

Insurers could make themselves more popular by recognizing that dynamic risk management could become a means for encouraging behavior change. Rather than offering non-negotiable premiums based on coarse models, the use of big data to assess individual risk would urge those customers to behave more responsibly. In this way, insurance could provide a price signal to nudge customers toward a lower-risk lifestyle. Insurers such as U.S.-based PruHealth have a healthy living rewards program, known as Vitality, which gives points for healthy activities such as regular gym attendance and not smoking. Points earned from the rewards program can also be redeemed for other lifestyle rewards such as cinema tickets or gift certificates.

Big data has the potential to create sophisticated risk models that are focused on individuals, extremely accurate, and capable of being updated in real-time. This is bad news for those hoping to use insurance as a means to justify excessive risk taking, but it is good news for those that want to be rewarded for managing risk more effectively. As more and more individuals opt for dynamic risk management, society will benefit from safer roads and smaller healthcare bills.

CATASTROPHE RISK

In 2006, Nicholas Stern alarmed the world. An economist by training, he postulated that extreme weather will cost the world one percent of the global gross domestic product (GDP) by the middle of the 21st century. In 2013, he said he had understated his original estimate.

Modeling Risk

Catastrophe risk models are computer-based approaches to measuring potential losses because of natural catastrophes such as earthquakes, windstorms, and floods. In the late 1980s, innovations in information technology and geographic information systems produced estimates of catastrophe losses by overlaying the properties in an insurer's portfolio with the potential natural hazard sources in the geographic area. After Hurricane Andrew made landfall in 1992, generating losses of $15.5 billion, the insurance industry quickly embraced catastrophe risk modeling as a new paradigm for doing business. The rapid adoption of catastrophe models also happened because of the fact that one of the original vendors, AIR World-wide, correctly estimated in real-time that losses from Hurricane Andrew could exceed $13 billion.

Catastrophe risk-management models assess a variety of risks, layered over geographic location. Geography is important, as location can determine certain types of risk.

Three of the major catastrophe risk-modeling vendors are RMS, AIR Worldwide, and EQECAT. At its core, insurance is about pricing uncertainty; hence, that is the business of these three vendors. Over time, their belief is that the models will get smarter, as more data is collected and overlaid with new data points.

OPEN ACCESS MODELING

Despite all of the success stories supporting catastrophe modeling, there are improvements to be made. The unexpected impact of the floods in Thailand during 2011 was not adequately modeled when large losses of around $45 billion arose from disruptions to manufacturing supply chains. This knock-on effect of natural disasters and business interruption is particularly difficult to model. Modeling determines risk. Data improves models. To the extent that data can help reduce uncertainty and do a better job of pricing for it, it can have a dramatic impact.

Risk assessment is essentially a forecasting challenge, and it is well known within the field of forecasting that you're wise to seek the opinion of multiple forecasters before making a decision. From a modeling perspective, a number of opinions equates to using a collection of diverse models, and so-called ensemble forecasting techniques are now popular amongst many weather service providers. Unfortunately, catastrophe-modelling vendors are reluctant to share their commercial models, which are often described as "black boxes" in the industry. This lack of transparency and failure to allow a means of combining models goes against the spirit of scientific innovation. Luckily, there are some seeds of change.

The Global Earthquake Model (GEM) was initiated in 2009 by the Organisation for Economic Co-operation and Development (OECD) in recognition of the need for catastrophe modeling to be made available worldwide. The GEM promotes an open-source approach to scientific research, data collection, and modeling. The GEM highlighted a need for greater transparency in estimating the risks associated with natural disasters.

In early 2014, a consortium of 21 insurers, reinsurers, and brokers unveiled the Oasis Loss Modeling Framework, which is a framework for independent catastrophe modeling. According to the consortium, the Oasis Loss Modeling Framework is "the most significant development in the modeling of natural catastrophe losses for 20 years." It is an open-access approach to sharing software, data standards, and methodologies for quantifying risk, for use by any party. Think of it as equivalent to an open marketplace for models and data: Anyone can contribute, anyone can use it.

OPPORTUNITIES

The total losses from disasters was $192 billion in 2013. When you start dealing with risks at that scale, the idea of using data to improve efficiency, reduce risk, and become more competitive is very attractive.

The following is a set of challenges and opportunities that the insurance industry faces today when incorporating big data:

- **Computation:** It is a technical challenge to navigate through a large corpus of data.
- **Interpretation:** There is a need to quantify uncertainty in the data and use sentiment analysis to uncover trends in unstructured data. Collaboration with third parties, brings a different perspective, typically leading to more complete interpretation.
- **Transparency:** There is a gap between quantitative analysis (data and models), and how the quantitative outcomes are deployed by policy-makers; therefore this gap has to be closed or narrowed.
- **Prediction:** Evidence-based policies require standards and protocols. Varying levels of confidence in predictions should be clearly communicated.
- **Scenarios:** Decision support, based on probability, has the ability to increase competitiveness and reduce risk.

The range of catastrophic risks in the world are significant and continue to evolve and emerge, elevating the business challenges and opportunities. With globalization and increased connectivity through social media, new risks emerge, threatening to lead to unexpected outcomes:

- **Interconnectivity:** Weather, food, water, energy, and politics are all intimately connected by human interactions.
- **Supply-chain risk:** Outsourcing and dependencies, throughout the flow of goods.
- **Reputational risk:** Working conditions in certain countries, as well as tax laws and perceived evasion.
- **Local versus global:** Local models have proven to be insufficient for calibrating potential global impacts, often leading to more risk than perceived.

So, where does the era of Big Data fit into this picture? The timeless equation for risk is a good starting point:

Risk = Hazard × Exposure × Vulnerability

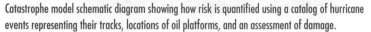

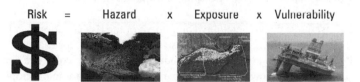

Catastrophe model schematic diagram showing how risk is quantified using a catalog of hurricane events representing their tracks, locations of oil platforms, and an assessment of damage.

In a catastrophe model, risk is quantified as the expected financial loss measured in dollars, hazard refers to the peril under consideration (e.g. hurricane, flood or earthquake), exposure provides details of the geographical locations of assets, and vulnerability quantifies the extent of the damage. The figure above illustrates the process for quantifying the risk of hurricane damage to oil platforms in the Gulf of Mexico. This type of catastrophe model equation embodies each of the emerging risks, challenges, and opportunities of managing these risks, and ultimately it's only as accurate as the data that feeds into the equation. Utilizing this equation puts an impetus on quantitative predictive modeling (augmented with qualitative judgment). Big-data approaches will support the fusion of scientific knowledge and empirical investigations. The goal is to leverage the equation to provide predictive analysis of risk, as opposed to retrospective analysis of risk. That can be the difference between insight and history. As Warren Buffet once said, "If history made you rich, librarians would be billionaires."

Approaches for dealing with time-varying hazards because of climate change is a good example of prospective risk modeling. Historical hurricane records tell us about the past, but the future may be different. Dynamical atmospheric models, similar to climate and numerical weather prediction models, are now capable of producing synthetic hurricanes as outputs. By introducing different climate scenarios, it is possible to generate a synthetic catalogue of hurricanes that can then be used to assess the potential losses. G. Anastasiades and P.E. McSharry (*Extreme value analysis for estimating 50 year return wind speeds from reanalysis data* [Wind Energy, 2014]) have developed an approach for using data generated by atmospheric models to provide sufficiently long-term records to assess windstorm risk to infrastructure, such as wind farms, where only short-term wind-speed records exist.

Big data is causing a revolution in insurance as a result of the many different sources of information that can be utilized to improve risk models. Satellite imagery determines the exact location and size of physical assets. The Internet and social media also provide a mechanism for allowing people to help with the collection of data necessary for quantifying risks.

In many cases, it is the vulnerability component of the risk equation that is most inadequate. Information about the structure of houses, for example, must be obtained from non-traditional sources. This form of collective data gathering, known as crowd-sourcing, is particularly important in developing countries where often no risk models exist.

MIDDLE OF SOMEWHERE, 2012 (CONTINUED)

As expected, inertia to change was to be found everywhere within the large multi-national insurer. Business processes had been forged with time and experience, and the desire to change was muted. However, willingness to change and knowledge that change is needed are two very different things. This company knew things needed to change; the question was whether or not they could summon the impetus to change.

In speaking with Chris Lewin, he was clear: "The work that actuaries do is bound to change in the future." As they become more data-centric, the challenge for actuaries will be to find new ways to bring data under control. "Taking a large mass of information (multi-dimensional) and turning it into much simpler summaries that enable the contemplative mind to understand data is no simple task," says Lewin. Large and complex data sets put the impetus on having the appropriate set of tools to aid the actuary in her craft. It's about tools, but it's also about culture, as the actuary and the company must be willing to evolve with or lead the new order in the industry.

In the case of this multi-national insurer, the insight to change came with their realization that they are sitting on top of a wealth of data, more than they could ever expect to utilize effectively. It seems that the complexity and size of the data was paralyzing to the organization. They are paralyzed by fear, instead of embracing opportunity.

4

PERSONALIZING RETAIL AND FASHION

KAROLINA

KAROLINA ZMARLAK WAS born in Poland during the Communist regime. Growing up in a Communist country teaches a person to make the best use of what little is available. Karolina remembers seeing her mother salvage scarce fabrics to have clothes made for her. This minimalism by necessity creates an enduring sense of inspiration.

Karolina immigrated to the United States in 1992, when her parents won a Clinton-era Green Card Lottery. She discovered her passion for sewing in high school, and she pursued her passion, graduating with high honors from the Fashion Institute of Technology in New York. Karolina is inspired by the notion that the right clothes, with the right fit, can change how a woman feels about herself. "When women wear my designs, I want them to feel empowered. I want my pieces to bring confidence," says Zmarlak.

While much of fashion has become de-personalized and mass-marketed, for Karolina, it's very personal. It is her craft, distinguished by attention to detail. For Karolina, the stitching and location of the stitching is just as important as the color or style of a particular item. Known as a ready-to-wear designer, she designs clothes to be worn with little alteration, but with the look of a custom fit. She refuses to produce them overseas, as an even 1 percent degradation in quality would fail the women that she seeks to empower. Accordingly, she is one of the few designers whose creations are fully designed, produced, and manufactured in New York City. It's not the cheapest place in the world to produce clothing, but her goal is quality and attention to detail.

Because of the established infrastructure in the fashion and retail industries, she is forced to seek traditional avenues to reach her clients: trunk shows, boutique retailers, department stores, web advertising, and presence at New York's annual Fashion Week. However, because of the high-end and craft nature of her designs, Karolina has a unique challenge: finding her clients. Or, said another way, helping her clients discover her.

Karolina believes she has 100,000 clients globally, each of whom will spend approximately $5,000 to $10,000 a year on her clothes. The problem is that many of her potential clients don't know that she exists. Data can serve as the matchmaker between Karolina and the women who are longing for her designs.

A BRIEF HISTORY OF RETAIL

Retail has continually reinvented itself over the past 100-plus years. Every 20 to 30 years, the form of retail has changed to meet the changing tastes of the public.

RETAIL ERAS

McKinsey & Company, the global strategy consultancy, has explored the history of retail in depth, citing five distinct timeframes:

- **1900s:** The local corner store was prominent in many towns. These small variety stores offered a range of items, including food, clothes, tools, and other necessities. The primary goal was to offer anything a person would need for day-to-day life.
- **1920–1940:** The corner store was still prominent but had grown to a much larger scale. In this era, department stores first began to emerge, and some specialization of stores began to occur.
- **1940–1970:** In order to effectively deal with some of the specialization seen in the previous era, this timeframe was marked by the emergence

of malls and shopping centers. This allowed for concentration of merchants, many of whom served a unique purpose.

- **1970–1990:** Perhaps best described as the Walmart era — a time when large players emerged, putting pressure on local store owners. These massive stores offered one-stop shopping and previously unseen value in terms of pricing and promotions. The size of these stores gave them economies of scale, which enabled aggressive pricing, with the savings passed on to the consumer.

- **1990–2008:** This era was marked by increased focus on discounting and large selection, coupled with the emergence of e-commerce.

Each era represented a significant innovation in the business model, but more important was the impact it had on each part of the retail value chain: merchandise and pricing, store experience, and the approach to marketing.

Each new era has longed for balancing the new innovations and expansion with a key hallmark of the past: customer intimacy.

ARISTIDE BOUCICAUT

Aristide Boucicaut was born the son of a hat maker. If you've ever walked into a hat store in Europe, it is an intimate experience and is often a walk through history. The different styles, from different eras, all tell a different story. Typically, there are used hats, new hats, and replicas of a foregone era. This notion of intimacy was part of Boucicaut's formative years.

In 1852, Boucicaut had the foresight to create what would become the first department store. He saw the opportunity to combine the intimacy of a hat maker's store with the selection and variety of a market. The creation of Le Bon Marché ushered in a new era of retail and commerce, with all the makings of a modern department store but careful attention to customer intimacy. Le Bon Marché marked the first time that consumers benefited from fixed prices, home delivery, and the ability to return items. In a sense, this store changed retail shopping and became the model for what retail shopping would become. As the store now says, "Le Bon Marché evolved into a resolutely Parisian, upmarket store whose values espousing authenticity and culture were closely bound with the joys of shopping."

Despite this unique combination of art and commerce, the concept of department stores is slowly losing its luster. A key contributor to the loss of relevance lies in the fundamental changes in retail: merchandise and pricing, store experience, and marketing. But there is something more happening here. The shift in retail has required the sacrifice of intimacy. And intimacy is what made it unique in the first place.

THE SHIFT

Amazon.com was launched in 1995. What started as the world's largest online bookstore has evolved into the world's largest store (online or otherwise). Although Amazon.com is not the sole culprit of the shift in retail, it is certainly an expediter of the shift and accordingly one of the most talked about.

In the last ten years, we have seen the bankruptcy of Circuit City, Linens & Things, Borders, CompUSA, Tower Records, Blockbuster, and countless other smaller players. The aforementioned big-box stores were designed to have the scale to withstand competition, yet in many cases, they declined at a faster rate than many regional and local stores.

A number of factors contribute to the decline in traditional retail:

- **Financial model:** Physical retailers have very high operating leverage, and therefore, if sales slow down at all, they are quickly short of the capital needed to fund the operation. E-commerce companies are much more efficient in terms of labor, real estate, and other factors, which accounts for the greater efficiency of the e-commerce financial model.
- **Price transparency:** Both online and offline commerce has created previously unseen pricing transparency, leaving little opportunity for profit margin expansion on products that can be found elsewhere.
- **Social engagement:** Social media, online reviews, and other interactive forums have placed an emphasis on price, quality, and experience. Nothing can be sacrificed, and everything is very public.
- **New distribution models:** Companies such as Stella & Dot are using a network of representatives, instead of physical stores, to distribute products.
- **Price of innovation:** The price of innovation has declined dramatically, enabling virtually anyone to enter a product category with something new and innovative. Previously, any type of product innovation would have required years of investment to get to market.
- **Experience:** In their book *The Future of Competition: Co-Creating Unique Value With Customers* (Harvard Business Review Press, 2004), C.K. Prahalad and V.J. Ramaswamy observed that value is increasingly co-created by the firm (in this case, retailer) and the customer, rather than being created entirely inside the firm. This flips the traditional model on its head; experience is no longer just about what the retailer offers, but is instead dominated by the influence of the consumer and the role that he or she plays in the experience. The authors state, "These personalized co-creation experiences are the source of unique value for consumers and companies alike."

In addition to the challenge posed by each of the factors in the preceding list, a retailer today faces extensive logistical challenges to manage the operation, while trying to do it at an appropriate level of profitability. Stores require warehouses, transportation, physical security, staffing at all links in the supply chain, and insurance at every point. Even worse, once you are inside the store, each store will have an extensive fixed-cost structure, including staff, inventory, rent, utilities, etc. It's a never-ending cost center, all in pursuit of intimacy. This is why revenue per employee for Amazon.com is $0.9 million per year versus $0.2 million at Walmart. Amazon.com simply has a better model.

These factors, when viewed together, clearly delineate the challenge facing traditional retailers, which has given rise to the wave of e-commerce. According to the U.S. Census Bureau, e-commerce grew to 6 percent of total retail in 2013.

Given these headwinds, the logical question is why we still have traditional retail at all? This question brings us back to Aristide Boucicaut and his passion for customer experience and intimacy. To date, e-commerce companies have struggled to achieve intimacy via their business model. But the Data era promises to change this deficiency.

THE DATA ERA

Retail, by definition, is mass market. It has been through every era. While subtle changes in approach have occurred, very few have captured the intimacy of the original corner store. The corner store's owner knew the customers personally; he understood what was happening in their lives, and the store became an extension of the community. In the Data era, mass marketing can reclaim the corner-store experience.

STITCH FIX

Stitch Fix is a data era retailer, focused on personalizing a shopping experience for women. While many women love clothes shopping, Stitch Fix realized that it is an inefficient experience today. It requires visiting many stores, selecting items to try on, and repeating. In fact, a successful shopping trip requires a relatively perfect set of variables to align:

- **Location:** A store must be near the shopper.
- **Store:** The store itself must interest the shopper and draw them in.
- **Clothing:** The clothing in the store must be of interest to the shopper.
- **Circumstance:** The clothing must match the circumstance for which the shopper needs clothes (dinner party, wedding, outing, etc.).

- **Size:** Even if all the preceding elements are present, the store must have the right size clothing in stock.
- **Price:** Even if all the preceding elements exist, the shopper must be able to afford the clothing.

To some extent, it's amazing that all of these variables ever align. And perhaps they do not, which leads to compromise. But if all the variables could align and occurred repeatedly, would the shopper be more inclined to buy? Yes, and hence the premise of Stitch Fix.

Stitch Fix is disrupting fashion and retail, targeting professional women shoppers who want all the variables to align. These women do not have the time nor perhaps inclination to search for the alignment and hence, Katrina Lake, the CEO and cofounder states, "We've created a way to provide scalable curation. We combine data analytics and retail in the same system."

When a person signs up for the service, she provides a profile of her preferences: style, size, profession, budget, etc. The data from that profile become attributes in Stitch Fix's systems, which promptly schedule the dates to receive the clothes, assign a stylist based on best fit, and enable the stylist to see the person's profile (meaning her likes and dislikes). The customer also specifies when and how often she wants to receive a *fix*, which is a customized selection of five clothing items. Then the data-and-algorithms team will present suggestions to the stylist. This recommendation system helps the stylist make great decisions. Once the customer receives the fix, she can keep what she wants and send back the rest. Stitch Fix obviously maintains the data on preferences so that, over time, it becomes a giant analytics platform, where recommendations can be catered to a unique shopper. Not since the corner store has such intimacy been available, and it's all because of the data. Clients are happier, the job of the stylist is easier, and this data then feed into the backend processes.

Retail is a difficult business. Fashion retail is even harder. It's not as simple as managing the supply chain (although that's not simple) because changing styles, seasons, and tastes are overlaid against the more traditional issues of sizes and stock. Any one poor decision can destroy the profit of a fashion retailer for a particular period, and therefore making the right decisions is at a premium. Stitch Fix attacks this challenge with human capital. Said a different way, this is not your typical management team for a fashion retailer. The leader of Operations at Stitch Fix comes from Walmart.com, while the analytics leader was previously an executive at Netflix. In a sense, Stitch Fix is building a supply chain and data analytics company that happens to focus on fashion. Not the other way around.

The company is making the bet that better customer insight will resolve many of the common fashion retailer issues: returns (ensuring fewer returns), inventory (predicting what people will want), and higher inventory turns (stocking things that customers will buy in the near-term). While Stitch Fix may not succeed as a retailer (although we think it will), it is laying the groundwork for the architecture of a retailer in the Data era.

Ms. Lake makes it clear that the company is first and foremost a retailer, but a retailer with a unique business model incorporating data and technology. Lake says, "We are a retailer. We just use data to be better at the core functions of retail. It's hard to buy inventory accurately without knowing your customer, so we use data in the sourcing process as well." She cites the example of looking at not just basic sizes (S, M, L or 2, 4, 6) as most buyers would, but looking at the detail of inseam size too. They can use this level of granularity in the buying process because of data. This attention to detail leads to a better fit for their clients and a higher likelihood those clients will buy.

Most data leveraged by Stitch Fix is generated by the company. Their advantage comes from the large amount of what Lake calls *explicit data,* which is direct feedback from clients on every fix. That's specific, unique, and real-time feedback that can be incorporated into future fixes and purchases. The buyers at Stitch Fix, responsible for stocking inventory according to new trends and feedback, love this data, as it tells them what to buy and focus on. As Lake says, "What customers buy and why, and what they don't buy and why not, is very powerful."

Stitch Fix has analyzed over 500 million individual data points. While the company has shipped over 100,000 fixes, no two have ever been the same. That's personalization. The company sells 90 percent of the inventory that it buys each month at full price, again because of personalization. Data and personalization have the impact of delighting clients while revolutionizing the metrics of retail.

KEATON ROW

Keaton Row was founded in 2011 and has raised over $2.5 million in venture capital funding, with the promise of delivering personalization and intimacy to women's fashion. It's long been known that peer pressure — or perhaps peer encouragement — drives the behavior of women shoppers. Said another way, if two friends are shopping together, they are much more likely to buy something with encouragement or advice from the other. Keaton Row duplicates this experience by pairing every customer with a personal shopper/stylist.

Elenor Mak, the cofounder and CEO of Keaton Row, learned the industry in her time at Avon, a beauty company, where she managed a large organization of sales representatives in the New York area. The lesson from Avon is about intimacy and connection. If two women learn about each other and have a vested interested in helping each other, they are more likely to listen to each other's advice. This makes Keaton Row stylists a digital version of the traditional Avon saleswomen.

A shopper who visits the website can quickly fill out a profile of her tastes in order to be matched with a personal stylist. While other e-commerce sites rely on statistical analyses and behavioral analysis (web click-throughs), Keaton Row is betting that human stylists will provide better advice and, more importantly, will provide the connection that compels women to purchase. "The Keaton Row customer is a professionally-oriented woman. She has money to spend, but doesn't have time. She isn't an active reader of *Vogue* or *The Coveteur,* so she wants it to be curated and convenient. There's a higher level of personalization, and authentic personalization," cofounder Cheryl Han said.

ZARA

Zara's business model is based on scarcity. In a store, if a shopper sees a pair of pants he likes, in his size, he knows it's the only one that will ever be available, which drives him to purchase impulsively and with conviction. Scarcity is a powerful motivator. In 2012, Inditex (the parent company of Zara) reported total sales of $20.7 billion, with Zara representing 66 percent of total sales (or $13.6 billion), with 120 stores worldwide. Scarcity can also be a revolutionary business model and profit producer.

Amancio Ortega was born in Spain in 1936. In 1972, he founded Confecciones Goa to sell quilted bathrobes. He quickly learned the complexity of fashion, extending to retail, as he operated this supply chain of his own creating. Using sewing cooperatives, Ortega relied on thousands of local women to produce the bathrobes. This was the most cost-effective way for him to produce robes, but it came with the complexity of managing literally thousands of suppliers. This experience taught Ortega the importance of vertical integration or, said another way, the value of owning every step of the value chain. He founded Zara in 1975, with this understanding.

Zara uses data to expedite the entire process of the value chain. While it takes a typical retailer 9 to 12 months to go from concept to store shelf, Zara can do the same in approximately two weeks. This reduced timetable is accomplished through the use of data: The stores directly feed the design team with real-time behavioral data. Zara's designers create approximately 40,000 new designs annually, from which 10,000 are selected for production. Given the range of sizes and colors, this variety of choice leads to approximately 300,000 new stock keeping units (SKUs) every year.

Zara's approach to the business has become known as fast fashion, as they will quickly adapt their designs to what is happening on the store floor, usher new products quickly to market, and just as swiftly move onto the next thing. This fast pace drives incredible efficiency in the implementation of the business model, yet at the same time, it creates enormous customer loyalty and intimacy, given the role of scarcity. Since the business can react so quickly, there is always sufficient capacity to produce the right design at the right time.

Zara's system depends on the frequent sharing and exchange of data throughout the supply chain. Customers, store managers, designers, production staff, buyers, and warehouse managers are all connected by data and react accordingly. Data drives the business model, but it's the *reaction* to the data that produces competitive advantage. Many businesses have a lot of data, but very few utilize it to rapidly effect decision making.

Unsold items account for less than 10 percent of Zara's stock, compared with the industry average of 17 to 20 percent. This is the data in action. According to *Forbes,* "Zara's success proves the theory that if a retailer can forecast demand accurately, far enough in advance, it can enable mass production under push control and lead to well managed inventories, lower markdowns, higher profitability (gross margins), and value creation for shareholders in the short- and long-term."

KAROLINA (CONTINUED)

Every time a woman eases into a Karolina style, her confidence is transformed. She can immediately sense the quality and style of Karolina's craft and is compelled to own something. More than half of the women who try something on eventually become clients, and then become repeat clients. It's the power of a great product, coupled with the right person.

In just three years of existence, Karolina has built a tremendous following. Her clothes have become a part of pop culture, appearing on *Saturday Night Live* and *The Today Show.* The women who wear Karolina know that they act and feel differently when they are wearing her clothing. They like the empowerment and confidence, and most of all, they love the fact that they love how they look.

Karolina will not achieve her potential as a designer through traditional retail outlets and approaches. Keaton Row, Stitch Fix, and Zara all provide a glimpse into the future, in terms of the power that data can have to transform fashion and retail. However, knowing what needs to be done is easier than actually doing it — therein lies the challenge for all fashion designers and retailers.

5

TRANSFORMING CUSTOMER RELATIONSHIPS WITH DATA

BUYING A HOUSE

A FRIEND WALKED into a bank in a small town in Connecticut. As frequently portrayed in movies, the benefit of living in a small town is that you see many people that you know around town and often have a first name relationship with local merchants. It's very personal and something that many equate to the New England charm of a town like New Canaan. As this friend, let us call him Dan, entered the bank, it was the normal greetings by name, discussion of the recent town fair, and a brief reflection on the weekend's Little League games.

Dan was in the market for a home. Having lived in the town for over ten years, he wanted to upsize a bit, given that his family was now 20-percent larger than when he bought the original home. After a few months of monitoring the real estate listings, working with a local agent (whom he knew from his first home purchase), Dan and his wife settled on the ideal house for their next home. Dan's trip to the bank was all business, as he needed a mortgage (much smaller than the one on his original home) to finance the purchase of the new home.

The interaction started as you may expect: "Dan, we need you to fill out some paperwork for us and we'll be able to help you." Dan proceeded to write down everything that the bank already knew about him: his name, address, Social Security number, date of birth, employment history, previous mortgage experience, income level, and estimated net worth. There was nothing unusual about the questions except for the fact that the bank *already knew everything they were asking about.*

After he finished the paperwork, it shifted to an interview, and the bank representative began to ask some qualitative questions about Dan's situation and needs, and the mortgage type that he was looking for. The ever-increasing number of choices varied based on fixed versus variable interest rate, duration and amount of the loan, and other factors.

Approximately 60 minutes later, Dan exited the bank, uncertain of whether or not he would receive the loan. The bank knew Dan. The bank employees knew his wife and children by name, and they had seen all of his deposits and withdrawals over a ten-year period. They'd seen him make all of his mortgage payments on time. Yet the bank refused to acknowledge, through their actions, that they actually knew him.

BRIEF HISTORY OF CUSTOMER SERVICE

There was an era when customer support and service was dictated by what you told the person in front of you, whether that person was a storeowner, lender, or even an automotive dealer. It was then up to that person to make a judgment on your issue and either fix it or explain why it could not be fixed. That simpler time created a higher level of personal touch in the process, but then the telephone came along. The phone led to the emergence of call centers, which led to phone tree technology, which resulted in the decline in customer service.

CUSTOMER SERVICE OVER TIME

While technology has advanced exponentially since the 1800s, customer experience has not advanced as dramatically. While customer interaction has been streamlined and automated in many cases, it is debatable whether or not those cost-focused activities have engendered customer loyalty, which should be the ultimate goal.

The following list identifies the main historical influences on customer service. Each era has seen technological advances and along with that, enhanced interaction with customers.

- **Pre-1870:** In this era, customer interaction was a face-to-face experience. If a customer had an issue, he would go directly to the merchant and explain the situation. While this is not scientific, it seems that overall customer satisfaction was higher in this era than others for the simple fact that people treat a person in front of them with more care and attention than they would a person once or twice removed.

- **1876:** The telephone is invented. While the telephone did not replace the face-to-face era immediately, it laid the groundwork for a revolution that would continue until the next major revolution: the Internet.

- **1890s:** The telephone switchboard was invented. Originally, phones worked only point-to-point, which is why phones were sold in pairs. The invention of the switchboard opened up the ability to communicate one-to-many. This meant that customers could dial a switchboard and then be directly connected to the merchant they purchased from or to their local bank.

- **1960s:** Call centers first emerged in the 1960s, primarily a product of larger companies that saw a need to centralize a function to manage and solve customer inquiries. This was more cost effective than previous approaches, and perhaps more importantly, it enabled a company to train specialists to handle customer calls in a consistent manner. Touch-tone dialing (1963) and 1-800 numbers (1967) fed the productivity and usage of call centers.

- **1970s:** Interactive Voice Response (IVR) technology was introduced into call centers to assist with routing and to offer the promise of better problem resolution. Technology for call routing and phone trees improved into the 1980s, but it is not something that ever engendered a positive experience.

- **1980s:** For the first time, companies began to outsource the call-center function. The belief was that if you could pay someone else to offer this service and it would get done at a lower price, then it was better. While this did not pick up steam until the 1990s, this era marked the first big move to outsourcing, and particularly outsourcing overseas, to lower-cost locations.

- **1990s to present:** This era, marked by the emergence of the Internet, has seen the most dramatic technology innovation, yet it's debatable whether or not customer experience has improved at a comparable pace. The Internet brought help desks, live chat support, social media support, and the widespread use of customer relationship management (CRM) and call-center software.

Despite all of this progress and developing technology through the years, it still seems like something is missing. Even the personal, face-to-face channel (think about Dan and his local bank) is unable to appropriately service a customer that the employees know (but pretend not to, when it comes to making business decisions).

While we have seen considerable progress in customer support since the 1800s, the lack of data in those times prevented the intimate customer experience that many longed for. It's educational to explore a couple pre-data era examples of customer service, to understand the strengths and limitations of customer service prior to the data era.

BOEING

The United States entered World War I on April 6, 1917. The U.S. Navy quickly became interested in Boeing's Model C seaplane. The seaplane was the first "all-Boeing" design and utilized either single or dual pontoons for water landing. The seaplane promised agility and flexibility, which were features that the Navy felt would be critical to managing the highly complex environment of a war zone. Since Boeing conducted all of the testing of the seaplane in Pensacola, Florida, this forced the company to deconstruct the planes, ship them to the west coast of the United States (by rail). Then, in the process, they had to decide whether or not to send an engineer and pilot, along with the spare parts, in order to ensure the customer's success. This is the pinnacle of customer service: knowing your customers, responding to their needs, and delivering what is required, where it is required. Said another way, the purchase (or prospect of purchase) of the product *assumed* customer service.

The Boeing Company and the Douglas Aircraft Company, which would later merge, led the country in airplane innovation. As Boeing expanded after the war years, the business grew to include much more than just manufacturing, with the advent of airmail contracts and a commercial flight operation known as United Air Lines. Each of these expansions led to more opportunities, namely around a training school, to provide United Air Lines an endless supply of skilled pilots.

In 1936, Boeing founded its Service Unit. As you might expect, the first head of the unit was an engineer (Wellwood Beall). After all, the mission of the unit was expertise, so a top engineer was the right person for the job. As Boeing expanded overseas, Beall decided he needed to establish a division focused on airplane maintenance and training the Chinese, as China had emerged as a top growth area.

When World War II came along, Boeing quickly dedicated resources to training, spare parts, and maintaining fleets in the conflict. A steady stream of Boeing and Douglas field representatives began flowing to battlefronts on several continents to support their companies' respective aircraft. Boeing put field representatives on the front lines to ensure that planes were operating and, equally importantly, to share information with the company engineers regarding needed design improvement.

Based on lessons learned from its first seven years in operation, the service unit reorganized in 1943, around four areas:

- Maintenance publications
- Field service
- Training
- Spare parts

To this day, that structure is still substantially intact. Part of Boeing's secret was a tight relationship between customer service technicians and the design engineers. This ensured that the Boeing product-development team was focused on the things that mattered most to their clients.

Despite the major changes in airplane technology over the years, the customer-support mission of Boeing has not wavered: "To assist the operators of Boeing planes to the greatest possible extent, delivering total satisfaction and lifetime support." While customer service and the related technology has changed dramatically through the years, the attributes of great customer service remains unchanged. We see many of these attributes in the Boeing example:

1. **Publications:** Sharing information, in the form of publications available to the customer base, allows customers to "help themselves."
2. **Teamwork:** The linkage between customer support and product development is critical to ensuring client satisfaction over a long period of time.
3. **Training:** Similar to the goal with publications, training makes your clients smarter, and therefore, they are less likely to have issues with the products or services provided.
4. **Field service:** Be where your clients are, helping them as it's needed.
5. **Spare parts:** If applicable, provide extra capabilities or parts needed to achieve the desired experience in the field.
6. **Multi-channel:** Establishing multiple channels enables the customer to ask for and receive assistance.
7. **Service extension:** Be prepared to extend service to areas previously unplanned for. In the case of Boeing, this was new geographies (China) and at unanticipated time durations (supporting spare parts for longer than expected).
8. **Personalization:** Know your customer and their needs, and personalize their interaction and engagement.

Successful customer service entails each of these aspects in some capacity. The varied forms of customer service depend largely on the industry and product, but also the role that data can play.

FINANCIAL SERVICES

There are a multitude of reasons why a financial services firm would want to invest in a call center: lower costs and consolidation; improved customer service, cross-selling, and extended geographical reach.

Financial services have a unique need for call centers and expertise in customer service, given that customer relationships are ultimately what they sell (the money is just a vehicle towards achieving the customer relationship). Five of the most prominent areas of financial services for call centers are:

- **Retail banking:** Supporting savings and checking accounts, along with multiple channels (online, branch, ATM, etc.)
- **Retail brokerage:** Advising and supporting clients on securities purchases, funds transfer, asset allocation, etc.
- **Credit cards:** Managing credit card balances, including disputes, limits, and payments
- **Insurance:** Claims underwriting and processing, and related status inquiries
- **Lending:** Advising and supporting clients on securities purchases, funds transfer, asset allocation, etc.
- **Consumer lending:** A secured or unsecured loan with fixed terms issued by a bank or financing company. This includes mortgages, automobile loans, etc.

Consumer lending is perhaps the most interesting financial-services area to explore from the perspective of big data, as it involves more than just responding to customer inquiries. It involves the decision to lend in the first place, which sets off all future interactions with the consumer.

There are many types of lending that fall into the domain of consumer lending, including credit cards, home equity loans, mortgages, and financing for cars, appliances, and boats, among many other possible items, many of which are deemed to have a finite life.

Consumer lending can be secured or unsecured. This is largely determined by the feasibility of securing the loan (it's easy to secure an auto loan with the auto, but it's not so easy to secure credit card loans without a tangible asset), as well as the parties' risk tolerance and specific objectives about the interest rate and the total cost of the loan. Unsecured loans obviously will tend to have higher returns (and risk) for the lender.

Ultimately, from the lender's perspective, the decision to lend or not to lend will be based on the lender's belief that she will get paid back, with the appropriate amount of interest.

A consumer-lending operation, and the customer service that would be required to manage the relationships, is extensive. Setting it up requires the consideration of many factors:

- **Call volumes:** Forecasting monthly, weekly, and hourly engagement
- **Staffing:** Calibrating on a monthly, weekly, and hourly basis, likely based on expected call volumes
- **Performance management:** Setting standards for performance with the staff, knowing that many situations will be unique
- **Location:** Deciding on a physical or virtual customer service operation, knowing that this decision impacts culture, cost, and performance

A survey of call center operations from 1997, conducted by Holliday, showed that 64 percent of the responding banks expected increased sales and cross sales, while only 48 percent saw an actual increase. Of the responding banks, 71 percent expected the call center to increase customer retention; however, only 53 percent said that it actually did.

The current approach to utilizing call centers is not working and ironically, has not changed much since 1997.

THE DATA ERA

Data will transform customer service, as data can be the key ingredient in each of the aspects of successful customer service. The lack of data or lack of use of data is preventing the personalization of customer service, which is the reason that it is not meeting expectations.

In the report, titled "Navigate The Future Of Customer Service" (Forrester, 2012), Kate Leggett highlights key areas that depend on the successful utilization of big data. These include: auditing the customer service ecosystem (technologies and processes supported across different communication channels); using surveys to better understand the needs of customers; and incorporating feedback loops by measuring the success of customer service interactions against cost and satisfaction goals.

AN AUTOMOBILE MANUFACTURER

Servicing automobiles post-sale requires a complex supply chain of information. In part, this is due to the number of parties involved. For example, a person who has an issue with his car is suddenly dependent on numerous parties to solve the problem: the service department, the dealership, the manufacturer, and the parts supplier (if applicable). That is four relatively independent parties, all trying to solve the problem, and typically pointing to someone else as being the cause of the issue.

This situation can be defined as a data problem. More specifically, the fact that each party had their own view of the problem in their own systems, which were not integrated, contributed to the issue. As any one party went to look for similar issues (i.e. queried the data), they received back only a limited view of the data available.

A logical solution to this problem is to enable the data to be searched across all parties and data silos, and then reinterpreted into a single answer. The challenge with this approach to using data is that it is very much a pull model, meaning that the person searching for an answer has to know what question to ask. If you don't know the cause of a problem, how can you possibly know what question to ask in order to fix it?

This problem necessitates data to be pushed from the disparate systems, based on the role of the person exploring and based on the class of the problem. Once the data is pushed to the customer service representatives, it transforms their role from question takers to solution providers. They have the data they need to immediately suggest solutions, options, or alternatives. All enabled by data.

ZENDESK

Mikkel Svane spent many years of his life implementing help-desk software. The complaints from that experience were etched in his mind: It's difficult to use, it's expensive, it does not integrate easily with other systems, and it's very hard to install. This frustration led to the founding of Zendesk.

As of December 2013, it is widely believed that Zendesk has over 20,000 enterprise clients. Zendesk was founded in 2007, and just seven short years later, it had a large following. Why? In short, it found a way to leverage data to transform customer service.

Zendesk asserts that bad customer service costs major economies around the world $338 billion annually. Even worse, they indicate that 82 percent of Americans report having stopped doing business with a company because of poor customer service. In the same vein as Boeing in World War II, this means that customer service is no longer an element of customer satisfaction; it is perhaps the sole determinant of customer satisfaction.

A simplistic description of Zendesk would highlight the fact that it is email, tweet, phone, chat, and search data, all integrated in one place and personalized for the customer of the moment. Mechanically, Zendesk is creating and tracking individual customer support tickets for every interaction. The interaction can come in any form (social media, email, phone, etc.) and therefore, any channel can kick off the creation of a support ticket. As the support ticket is created, a priority level is assigned, any related history is collated and attached, and it is routed to a specific customer-support person. But, what about the people who don't call or tweet, yet still have an issue?

Zendesk has also released a search analytics capability, which is programmed using sophisticated data modeling techniques to look for customer issues, instead of just waiting for the customer to contact the company. A key part of the founding philosophy of Zendesk was the realization that roughly 35 percent of consumers are *silent users,* who seek their own answers, instead of contacting customer support. On one hand, this is a great advantage for a company, as it reduces their cost of support. But it is fraught with risk of customer satisfaction issues, as a customer may decide to move to a competitor without the incumbent ever knowing they needed help.

Svane, like the executives at Boeing in the World War II era, sees customer service as a means to build relationships with customers, as opposed to a hindrance. He believes this perspective is starting to catch on more broadly. "What has happened over the last five or six years is that the notion of customer service has changed from just being this call center to something where you can create real, meaningful long-term relationships with your customers and think about it as a revenue center."

BUYING A HOUSE (CONTINUED)

It would be very easy for Dan to receive a loan and for the bank to underwrite that loan if the right data was available to make the decision. With the right data, the bank would know who he is, as well as his entire history with the bank, recent significant life changes, credit behavior, and many other factors. This data would be pushed to the bank representative as Dan walked in the door. When the representative asked, "How can I help you today?" and learned that Dan was in the market for a new home, the representative would simply say, "Let me show you what options are available to you." Dan could make a spot decision or choose to think about it, but either way, it would be as simple as purchasing groceries. That is the power of data, transforming customer service.

6

INTELLIGENT
MACHINES

DENMARK

AS I SAT in the cockpit of a Tesla roadster for the first time, I was struck by the compact feel of the car and the tight steering wheel. Applying the gas for the first time reminded me of the feel of a golf cart: Push the accelerator and you lurch forward. Take your foot off the accelerator and you immediately begin to slow. There is really no need for brakes. The car almost seems to anticipate every move from the driver, whether it is slowing into a curve or accelerating through a bend; the car simply responds.

It was 2009, and I was visiting Denmark for the first time in my life. Copenhagen? Nah. My first trip was to Aarhus, the second largest city in Denmark. It has one of the largest industrial harbors in Northern Europe, seeing 8,000 ships a year docking. Aarhus is a vibrant area of innovation around IT, life sciences, and nanotechnology, and is perhaps best known for research and innovation in the environment, energy, and science.

I was in Aarhus to visit Vestas, the world's largest wind turbine manufacturer. While Vestas has been challenged through the years by General Electric, Sinovel (in China), Siemens, and others, it always seems to retain a position at or very near the top of the rankings of wind turbine manufacturers. Talking with the engineers at Vestas, I quickly realized that this is not a job; it's a passion, a hobby, and in most cases, a love for clean energy and innovation that drives the team.

Wind energy has never really reached its potential, primarily due to the financial landscape that both the manufacturers and the customers face. Here are some of the attributes of a wind turbine business:

- **Cost and price:** Buying a wind turbine is typically $80 million to $100 million (its *price*), while that same turbine will *cost* around $65 million to $80 million to manufacture. This sales transaction represents a big capital purchase, without that much profit for the manufacturer.
- **Useful life:** Despite the significant investment, the useful life of a turbine is in the range of 20 to 30 years, with some variability based on the size and type of turbine. The average turbine has blades of around 150 feet in length and some turbines' rotor diameters stretch between 250-430 feet.
- **Productivity:** Turbines are useful only if you are able to transport the energy generated, which means that special consideration has to go into where it is located.
- **Location:** Since they are so large, most people do not want one in their backyard, which means that remote locations become common. Off-shoring turbines has become more common as of late, but that makes it even more difficult to transport the energy that is generated.
- **Service:** Inevitably, these complex machines will need to be serviced, which gets complicated when you consider location (sometimes they are offshore) and scale (it's not easy to get to the top of a turbine).

The promise of turbines becoming more intelligent addresses many of the difficulties that are inherent in the business. If the machines can service themselves and consciously extend their useful life automatically, then those two factors alone can dramatically alter the economics. In some cases, turbines have been called dumb windmills, spoiling rural landscapes. However, with advances in technology, turbines are quickly becoming beacons of intelligence on the countryside, emanating data and intelligence.

INTELLIGENT MACHINES

In 1984, film director James Cameron first found success his science fiction hit *The Terminator*. He co-wrote the screenplay with the film's producer, and the plot, which seemed quite unrealistic when the film was released, seems a bit more relevant today. The story is about a cyborg assassin that arrives in the year 1984 (from the year 2029), programmed to kill a particular woman. The assassin arrived from a future where intelligent machines have taken over the human race, and the cyborg travels back in time to ensure that this never changes.

Cyborg assassins? Not so realistic. But the notion that machines will become so intelligent that they can outsmart humans is not as far off in the future as we may want to think. In 1997, IBM's computer Big Blue beat the world chess champion after a six-game match: two wins for IBM, one for the champion and three draws. This inspired researchers at IBM to design a machine that could beat the champions at a more complicated game, known as *Jeopardy!*. In 2011, IBM's computer, called Watson, beat two of the all-time most successful human players of the game.

MACHINE DATA

The intelligence of machines is based on their ability to generate, analyze, and act on data. There are many modern-day terms for the phenomenon around machine generated data: Internet of Things, industrial Internet, sensors and actuators. Each term is illustrates a world in which machines are collecting and analyzing data at scale, and then using that data to drive action.

Machine data includes all data that is generated by any machine. This could include any device or object that is connected to a network, such as sensors, servers, networks, applications, automobiles, turbines, and railroad cars. This is where the term *Internet of Things* came from, as it is encompassing anything that is connected to the Internet and is therefore a data point or a source of data on the overall network. The types of machine data can be diverse and far ranging, but some more common ones include:

- **Database logs:** Log files produced by an embedded database.
- **Application logs:** Data from any log file type from any application.
- **Application server logs:** Data from the integration of applications, including operational details and performance metrics.
- **Call detail records:** Logs generated from a telecom network, detailing each call and session.
- **Sensor data:** Data from a sensor on any object, ranging from a container on a ship to a railroad car. This data records the exact position of a device at a specific moment.
- **GPS data:** As discussed in Chapter 1, "Transforming Farms with Data," GPS data is a critical source of data, but with applications much beyond agriculture.

- **IP router data:** Offers insight on the performance of a network and the devices or objects that are interacting with that network.
- **Clickstream data:** Data generated from the clicks on a web page, which creates a stream of data on behavior and activity with that web page.
- **Transactional data:** Records of financial payments registered by credit and debit card purchases.

This is just a sampling of the types of machine data that can be collected, analyzed, and acted upon. Value can be derived from collecting and analyzing the data in its native form.

In 2009, The Gartner Group published a report projecting that data would grow by 650 percent over the next five years, with 80 percent of the data being unstructured in form. This projection was exceeded and most of that growth has come from machine data. IDC took it a step further with the prediction that, in 2020, there will be 26 times more connected things than people. While we are often captivated by the number of human-operated devices and smartphones, it is evident that this form of data production is quite minor when viewed in the context of the total number of data-producing objects. Machine data may prove to be the real frontier in the Data era.

THE DATA ERA

It is estimated that by 2020, 40 percent of all data in the world will be machine-to-machine in form. Airplanes are known to generate approximately 2.5 billion terabytes a year, while sensors on oil wells generate exabyte-size amounts of data. Intelligent machines, producing previously unimagined amounts of data flow, will transform industries and companies.

GENERAL ELECTRIC

In November 2011, General Electric announced a $1 billion investment in software. While GE attributes a small portion of its revenue to software historically, this announcement was significant in terms of its scale and the strategic focus. GE's revenue contribution has traditionally come from health-care (~20 percent), power and water (~40 percent), aviation (~20 percent), and finance/other industrials (~20 percent). Software is a relatively new frontier.

When you drive into GE's software facility in San Ramon, California, it is striking how dramatically different it is than what you would expect from GE. San Ramon, while technically in the San Francisco Bay Area, is a far cry from Silicon Valley. The spacious campus, perfectly manicured, has the feel of an East Coast headquarters, but with an environment that is not often found on the East Coast.

GE has decided to build a software team targeted at the opportunity in machine data or, as GE calls it, the Industrial Internet. GE sees the opportunity to mine the massive data sets that it generates every year, from aircraft engines to railroad cars to nuclear power plants. This data can be used to predict failures and improve efficiency, or possibly could be turned into a business on its own merits: selling data back to the clients who use the machines.

Industrial Internet

A year after the announcement of the software investment, Peter Evans and Marco Annunziata penned a 2012 paper titled "Industrial Internet: Pushing the Boundaries of Minds and Machines." While this paper is sponsored and written by General Electric, it contains some important insights that apply well beyond GE.

As we learned in the Prologue of this book, in the story about Dave Brailsford and the British cycling team, small gains in performance can have a huge impact on overall results. This phenomenon is particularly acute when it comes to impacting large capital equipment, where a simple 1 percent in fuel savings, for example, could drive $30 billion of value over a 15-year period.

The paper goes on to describe three distinct waves of innovation over the last 200 years:

- **Wave 1: Industrial Revolution:** The emergence of machines, equipment, and factories automating many tasks that were previously done by humans. This first brought scale to the production of many materials and products. This period lasted approximately 150 years.

- **Wave 2: Internet Revolution:** The rise of information networks, driven by connectivity to networks for people and businesses and a similar rise in computational power. As we've seen in the last 20 years, this wave has transformed many businesses.

- **Wave 3: Industrial Internet:** A new era that is dawning, in which machine data and analytics will automate many tasks that require human decision making today. In Waves 1 and 2, humans were necessary to ensure that machines and networks operated effectively and efficiently. In Wave 3, the machines will be intelligent enough to do this themselves.

Software will be the primary enabler of Wave 3. This software will not only analyze the data generated from machines, but also will provide the tools to manage the machines themselves. This is why GE had to make a big bet in software when many people believed it was perhaps non-core to their business. "What will all of this look like when 50 billion machines around the world get connected?" asks William Ruh, the head of GE's San Ramon operations. "It will certainly change the way we operate machines and how we operate the company. And it will change lots more besides."

DRONES

Drones have been used in the military, either as weapons or for surveillance, for years. In the 1990s, drones took center stage with the unveiling of the Predator drone during the Gulf War. At the time, they were referred to as Unmanned Aerial Vehicles (UAVs), but drones have applications far beyond the battlefield.

There has also been a fairly long running subculture of drone enthusiasts who build and fly them as a hobby. Behind the scenes, commercial usage of drones in energy, agriculture, and media has been on the rise but not really understood. The general public finally woke up to the business opportunity around these intelligent machines when Amazon.com, in its marketing genius, appeared on the television news show *60 Minutes* on December 1, 2013. In that segment, Amazon.com highlighted a research and development project to utilize drones for package delivery to homes. This well understood application (delivering packages) ushered drones into the mainstream, as the average person could now imagine a practical benefit. The incredible potential has been recognized by the launch of the Drones for Good Award by the United Arab Emirates, which offers a prize of $1 million for the best civilian use of drones for improving people's lives.

Energy

Oil is very hard to find, transport, and refine, primarily due to the remote locations that typically house profitable drilling locations. It is normal for a drilling location to be far from cities, remote from established infrastructure (i.e. without access to roads), and difficult to get to. Offshore rigs and barren Alaskan landscapes come to mind. These inhospitable locations make pipeline and drill-site inspection very difficult.

Many oil companies have discovered that drones with special sensors and cameras flying over pipelines become a very convenient and efficient way to detect leaks, monitor operations, and collect data. Unmanned aerial vehicles flying over pipelines, while outfitted with special sensors, could detect leaks quickly.

British Petroleum (BP) was the first company approved to use commercial drones to monitor pipelines, and more recently, ConocoPhillips conducted a test in Alaska's Chukchi Sea. Their objective was to survey environmental characteristics, recording pertinent data and changes over time, so as to ensure adherence to regulations.

Observing the utility of drones led Aeryon Labs to launch a business in Canada, offering a low-cost mechanism to inspect and manage energy sites (smokestacks, turbines, and rigs). Rather than hiring helicopters that can cost thousands of dollars per hour to operate, the company began to utilize drones. Aeryon has the unique advantage of operating in Canada, where drone regulations are much more relaxed than what the Federal Aviation Administration (FAA) allows in the United States.

As drones become more prominent, the key question for energy companies will be what they do with the data once a drone has collected it. Deriving value from new data sets is often easier said than done.

Skycatch

Skycatch, founded in 2013, has raised $19.6 million to pursue their ambition to commercialize the use and application of drones. Not surprisingly, Google, which has an unabashed interest in collecting data, is an early investor in the company.

Christian Sanz, Skycatch's CEO, started like many in this arena: as a hobby-ist. He began building drones and experimenting with flying ranges, and eventually, he put a camera onto one of his drones. For fun, he offered to fly it over a local construction site, taking pictures to document progress on the project — and sure enough, the construction managers could not get enough of the information. It wasn't the drones themselves that compelled Sanz to start the business. It was the immense interest in the data that could be collected from this new viewpoint.

Drones from Skycatch are the future of intelligent machines. When a drone completes a flight, it automatically returns to its station on the ground. It is guided in for its landing through the use of its camera, GPS, and other sensors. When the drone lands, the robotics in the base station automatically remove the battery (which stores 15 gigabytes of data), and refresh the battery. Then, the data from the removed battery is uploaded to a data-cloud service for processing and analysis. An entire process that would normally take multiple days and multiple humans is now done in minutes without intervention. That is the promise of intelligent machines.

The Market

Economics and data are driving the drone market. It is by far the cheapest and most efficient way to analyze remote areas or large sites, whether they are energy-related areas, construction zones, farms, or other sites. This efficient analysis creates previously untapped data sets for analysis and exploitation.

For this reason, the global drone market is expected to grow to between $10 billion and $30 billon over the next decade. The market size range is dramatic, due to uncertainties about regulations and consumer acceptance.

TESLA

Tesla has built its reputation as an innovator in electric automobiles. All of the attention that the company receives focuses on its co-founder and CEO (Elon Musk), its batteries and charging station penetration, and its battery factories (marketed as Gigafactories). Yet none of those factors are as compelling as the software innovation in and around Tesla that makes it a modern-day intelligent machine.

On a recent earnings call, Musk commented:

> *We continue to make software improvements. So I think there are some very exciting software updates. They're going to come out in the next few months that will improve the experience for the whole fleet of customers out there. . . . I think customers can certainly look forward to some really awesome functionality improvements in their existing car.*

Tesla has what it calls a Vehicle Management System onboard every car. It is the nervous system of the vehicle, managing the security, communicating any relevant messages, and enhancing the driver's experience, all made possible by the sensors and processors all over the vehicle, which feed streams of data into the Vehicle Management System. This data can be utilized to do things on behalf of the driver (battery management, motor diagnostics, door locks) and it can be utilized to work with the driver (share information, react to externalities, etc.). Lastly, it manages the driving modes and is constantly computing the range of the vehicle, based on data regarding battery age, capacity, and the form of driving.

The software in the vehicle is the intelligence behind the Vehicle Management System. Version 6 of the software extended the experience to include things like manual adjustment of suspension height and three-phase charging for European countries. Speculation indicates that in the future, it may include integration with third-party applications, intelligent cruise control, and the ability to render your smartphone screen on the display module. In addition, it is expected that future versions will incorporate deeper integration of GPS, whereby the car will constantly run calculations to optimize routing and drive times based on traffic, construction, or other factors. This data not only will come from the Tesla that a person is driving, but will be combined with the data coming from all other Teslas on the road.

To complement the intelligent vehicle, Tesla has built a customized back-end management system called Warp that combines commerce, supply-chain, and vehicle analytics into a single platform. Drivers can request service through Warp and receive software updates to fix their issues. In one instance, customers were complaining that the vehicle would roll back slightly when stopped on a hill at a red light or stop sign. This issue was reported through Warp, and then Warp provided a software update to address the issue. This is likely the first time in history that the braking system on a vehicle was adjusted without a single person looking at or touching the car; it was all via software and the Internet.

There is no disputing that Tesla is a number of years ahead of most vehicle manufacturers when it comes to intelligent machines, the use of data, and the application of software. But, in time, this type of intelligent machine will be standard.

NETWORKS OF DATA

Whether it is drones, Teslas, wind turbines, or an aircraft engine from General Electric, each of these intelligent machines becomes a node on a network of data. The intelligence lies not only in the machines, but also in the network itself. And the aggregation of data in the network can be funneled back to the machines (nodes) to improve their operation and productivity.

Networks of data are forming in many industries, even those in which you would least expect it. Even the shipping industry, which has been around since the 15th century, has begun to operate as a network of data. Cargo ships, containers, and the ports that they utilize have evolved to be nodes on a vast global network by leveraging GPS technology on ships, RFID tags on containers, and track/trace technology. A look at how the number of countries acting as nodes on the shipping network has evolved since 1960 is striking.

The increase in shipping nodes has altered trade balances and, perhaps more importantly, has altered the landscape of the shipping network. Now, the number of nodes is not just about the number of ships, containers, or ports; it's also the number of countries participating.

Similar networks of data are being formed in every industry and geography. Understanding these networks, and participating or creating them, will shape competitive advantage in the next century.

DENMARK (CONTINUED)

We are quickly ushering in an era in which wind turbines will be just as intelligent as the Tesla. They will be self-running and self-optimizing, and data will drive the decision making, instead of human insight. The turbine will no longer be a windmill in the ground but instead will act as a node on the network. That network could be as broad as the global energy grid or as narrow as a few nodes to collect data on weather conditions in a single location. Either way, intelligent machines and the networks of data that they create will alter the future of many companies and industries.

7

GOVERNMENT AND SOCIETY

EGYPT, 2011

IMAGES FROM TAHRIR square in Cairo, Egypt during early 2011 serve to remind us of the power of a coordinated group of young protestors and the threat that such a movement can represent for the ruling government. The "Arab Spring" is the name given to the protests, demonstrations and civil unrest that started in December 2011 and spread throughout the Arab world. While some observers compare the Arab Spring with the "Revolutions of 1989" in Eastern Europe, the main difference has been the use of social media as a means of communication for successfully organizing and coordinating these protests.

Tunisia, Yemen, Libya, and Egypt have all witnessed the fall of their rulers and a civil war is ongoing in Syria. While Tunisia is making good progress with establishing democracy, both Yemen and Libya are still struggling. Egypt's troubles continue as its newly elected president, Mohamed Morsi, was locked up in prison and violence is on the rise again.

SOCIAL MEDIA

Government is increasingly realizing that the Internet and social media offer a means of understanding society and designing policies that have a greater likelihood of being both accepted and effective. One likes to imagine that the government has our best interests at heart and that its intelligence agencies are dutiful in protecting us against threats to our way of life that might arise from terrorism, criminal activity, and cybercrime. Throughout history, governments and rulers have used intelligence as a means of keeping one step ahead of adversaries. The emphasis now should also be placed on improving the socio-economic wellbeing of all members of society. By focusing on the measurement of both rewards and risks for individuals and society at large, it will be possible to create a knowledge-based society where the policymaking process is transparent and supported by empirical evidence.

Data is central to gathering intelligence, and big data analytics is helping deliver the capability to turn this haystack of data into insight, early warning systems, and a basis for appropriate action. Since 2007, the ability to generate information outpaced the ability to store this information. Due to innovations in information and communications technology (ICT) around the Internet and mobile telephone, the transmission of emails, SMS, voice, music, and video is responsible for the production of an ever-increasing amount of data.

These developments herald a game changer in the relationship between government and society. No longer does government have to hold an expensive referendum to assess public opinion with regard to a crucial decision. Rather than asking citizens what they think, big data provides sufficient blogs, tweets, web searches, and transactions to assess opinions in real-time and to forecast likely responses. At present, it might be difficult to justify using data to design new policies, but over time and after thorough evaluation, this approach may become more common.

INTELLIGENCE

Throughout history, rulers have used a variety of methods to gather intelligence. Knowing what potential adversaries are doing and those adversaries' capabilities is obviously a substantial advantage. In the data era, it is apparent that intelligence agencies rely more and more on technology to collect and analyze data generated by human activity. The immense advantages of superior technology and intelligence are well known in business and can provide a competitive edge. The incredibly fast growth and success of Google is due to providing access to novel sources of data and an ability to create analytics for facilitating everyday activities such as searching, communicating, and collaborating.

SNOWDEN EFFECT

The whistleblower Edward Snowden wrote in a note accompanying the first set of documents he provided, "I can't in good conscience allow the U.S. government to destroy privacy, internet freedom, and basic liberties with this massive surveillance machine they are secretly building." This statement from Snowden changed the way many people view intelligence agencies. Whether based on Ian Fleming's character James Bond or John le Carré's books, secrecy and intrigue have always been an important part of working in intelligence. Snowden expressed a view that the agencies, in an attempt to monitor the increased threat of potential terrorists following 9/11, had gone too far by collecting information about a large swathe of society. Concerns were widespread that communications, activities, and transactions of innocent people were being analyzed with a fine-tooth comb. In addition, the accusations were that the intelligence agencies were exchanging information and collaborating to bypass national laws.

It was obvious that the repercussions of Snowden were having some impact when, for the first time in history, the chiefs of the three intelligence agencies in the U.K. (MI5, MI6, and GCHQ) volunteered to be interviewed by the House of Commons's Intelligence and Security Committee in November, 2013. In an attempt to reassure society, these chiefs explained that their organizations were subject to a high level of ministerial scrutiny and operated within the law. They also explained in their testimony that espionage against the U.K. was happening on an "industrial scale" and that Al-Qaeda was "lapping up" the security leaks given to the media by Edward Snowden. Unfortunately for these agencies, as their mission is to prevent terrorism, success is measured by avoiding news events, and therefore society is not aware of the impressive work that takes place nor of the number of lives that have been saved. The committee heard that 34 terrorist plots had been foiled since suicide bombers killed 52 people in the 7/7 blitz on London commuters in 2005.

PRIVACY RISK VERSUS REWARD

There is a delicate balance between the collective desire for security and the need to reduce privacy risk. Attitudes towards this balance are cultural and depend on history. Intelligence agencies may be welcomed by democratic societies but feared by those ruled by authoritarian regimes. While the citizens of democratic governments may believe that they have nothing to hide, these revelations by Snowden raised important questions about the power and utilization of data. While there is an expectation that the objective of intelligence agencies is to increase national security, fears about espionage in both the public and private sectors have spread.

Even when a society trusts its democratically elected government, there should be concerns about anonymity and the risk of data being accessed by other organizations that have less than benevolent motives. Following the revelations by Snowden, government agencies have been under increased scrutiny. In early June 2014, Germany opened an inquiry into claims that the U.S.'s National Security Agency (NSA) tapped the mobile telephone of the German Chancellor, Angela Merkel. A few days later, Vodafone released its first transparency report stating that a small number of governments have direct access to communications flowing over its networks.

There have already been a substantial number of cases in which data records held by a government have been compromised. In the U.K., two CDs that contained records for 25 million individuals, including details of the families of Child Benefit claimants, went missing in the mail. The U.K. Driving Standards Agency lost a hard drive containing details of 3 million candidates for the driving theory test. This suggests that greater importance should be given to securing and encrypting sensitive information about individuals.

Vulnerabilities on the websites of large corporations are just as worrying and often of greater scale. During March 2014, eBay's security was compromised when the online auction site was hit by a cyber-attack. After its database, containing sensitive information of 145 million people, was hacked into, it had to urge its clients to change their passwords. As cybercrime increases, it is well acknowledged that in the hands of criminals, stolen data records could pose a substantial threat.

It is all too easy to focus on the risks and for paranoia to escalate. The antics of government departments and failures of corporations to secure their information certainly provide good media content. Luckily, lost data does not always find its way into the hands of villains. The digital economy relies on our ability to move bits of data around in order to reduce transaction costs. It is hard to imagine moving back to an era of pushing paper around if the Internet and smartphones could not be used to purchase goods.

In Africa, mobile phones provide the infrastructure for many innovative financial services. Rather than playing catch-up with the developed world, Africa is leapfrogging ahead with mobile payments and mobile insurance. Without fearing the large costs of shedding legacy infrastructures, start-ups in Africa are incredibly nimble. Furthermore, the sluggish and traditionally conservative approaches of the financial sector are being given an energy boast and face-lift as innovative mobile network operators provide the back-bone of this revolution.

OBSERVATION OR SURVEILLANCE

Citizens in the developed world raise concerns about surveillance and actively voice opinions about invasion of privacy. Yet in the developing world, poor people are frequently observed as part of research projects. Usually with the best of intentions, development organizations test out different interventions in an attempt to improve the lives of those that need the most help. But just because someone is poor does not mean that their right to privacy should be ignored. Adequate approaches to protecting the privacy of individuals who provide data need to be developed. Again, it is important to recognize that setting standards for privacy in experiments in the data era will involve reaching a compromise between maximizing rewards and minimizing risk. As datasets are anonymized and geographical location is degraded, the chances of finding patterns that provide predictability and a mechanism for innovation diminishes.

DEVELOPMENT TARGETS

In a marked shift of approach, since the year 2000, international development has been increasingly driven by data, with the introduction of eight Millennium Development Goals (MDGs), which represent a blueprint agreed to by the United Nations and leading development institutions such as the World Bank, U.S. Agency for International Development (USAID) and the U.K. Department for International Development (DFID) and regional development banks. The aim is to achieve the following MDGs by 2015:

1. To eradicate extreme poverty and hunger
2. To achieve universal primary education
3. To promote gender equality and empower women
4. To reduce child mortality
5. To improve maternal health
6. To combat HIV/AIDS, malaria, and other diseases
7. To ensure environmental sustainability
8. To develop a global partnership for development

Progress toward meeting each of these goals is monitored using specific targets and data-driven indicators. For example, the first target for the first goal above is to halve, between 1990 and 2015, the proportion of people whose income is less than one dollar a day. This particular target is monitored using three indicators: the proportion of population below $1.25 (PPP) per day; the poverty gap ratio; and the share of poorest quintile in national consumption. In June 2005, the G8 Finance Ministers agreed to provide enough funds to the World Bank, the International Monetary Fund (IMF), and the African Development Bank (AfDB) to cancel $40 billion to $55 billion in debt owed by members of the Heavily Indebted Poor Countries (HIPC) to allow these countries to redirect resources to programs for improving health and education and for alleviating poverty.

With these eight goals defined with a deadline of 2015, the new debate has shifted to the post-2015 agenda and what should follow the MDGs. Despite the wide debate around the key aspects of this new agenda, it is broadly agreed that quantitative approaches and a greater use and acceptance of data are essential in both the design and the evaluation of development programs funded by international donors and aid agencies. Notwithstanding the arguments made by social scientists about the perverse behavior caused by using specific metrics, it is now widely acknowledged by the development community that having targets in place helps focus attention and increases the chance of success.

As the UN debates the post-2015 agenda, there has been repeated reference to making sure that this agenda is based on evidence, and there are growing pressures to link funding to proven impact. The UN's High Level Panel report coined the term *data revolution* to draw attention to the pressing need to ensure no one gets left behind as development occurs and issued a call for transparency and accountability for governments and development agencies. Exciting new methods, for collecting data using SMS surveys and evaluating projects using randomized controlled trials, provide solid foundations for understanding which development programs and interventions actually work.

Big data offers the potential of defining and selecting key performance indicators (KPIs) that are relevant for both the recipients of aid and the donors who fund the projects. SMS-based platforms can easily be deployed to facilitate surveys, and direct beneficiary feedback offers a means of monitoring progress in real-time. Indeed, recent work by Sachin Gathani, Maria Paula Gomez, Ricardo Sabates, Dimitri Stoelinga ("The Survey Effect: How data collection frequency boosts outcomes in a coffee agronomy training program in Rwanda," *Laterite Working Paper*, 2014) suggests that monitoring in and of itself can increase the likelihood of success. The work of Abhijit V. Banerjee and Esther Duflo ("Poor Economics: A Radical Rethinking of the Way to Fight Global Poverty," *Public Affairs*, 2011) has also highlighted the importance of using data in measuring the impact of development and taking a more rigorous approach to evaluation. Increasingly, data is moving from the world of geeks to become a key part of all policymakers' decision-making tool kit.

An initiative of the United Nations known as the UN Global Pulse is demonstrating the potential of big data by exploring innovative methods and frameworks for combining new types of digital data with traditional indicators to track global development in real-time. By forging strategic public-private partnerships, it hopes to secure access to sources of data, state-of-the-art analytical tools, and expert advisors in the relevant technical fields. With headquarters in New York and Pulse labs in Indonesia and Uganda, it focuses on applied research projects that address wide-ranging topics including food security, humanitarian logistics, economic well-being, gender equality, and health.

OPEN DATA

Data is unlikely to help governments unless they cooperate in making data openly available. Fearing that this could expose weaknesses, many governments have been reluctant to hand over the key to the data treasure trove. Others have encouraged some of their institutions to do so without thinking about the advantages and how to build innovation in partnership with those organizations that wish to utilize open data sets. Simply making data freely available is only a small step in the right direction. It is also necessary to think carefully about the design of web portals and platforms for providing this data in a clear and consistent structure so that it can be widely used by a broad spectrum of stakeholders. Without such structures, the portals resemble haystacks, and it is unlikely that external organizations will have the desire or patience to keep searching for a needle within these.

HACKATHONS

Hackathons are a phenomenon that has arisen at the interface of big data and computing enthusiasts. The general recipe is to provide a space with some food and drink and invite those interested in working intensely on a problem for a day or more. The most effective hackathons have a clear focus, carefully prepared structure, and group of end-users who will actually implement the solution. As the novelty of providing one's time for free at hackathons wears off, it becomes obvious that attending a hackathon is akin to a charitable offering, and therefore only those challenges that are truly aiming to provide benefits to society are likely to meet with success. For this reason, it is important that government takes an active role in organizing and supporting hackathons. By bringing novel data and a worthy challenge, government can ensure that the outputs are translated into practical solutions.

OPEN ACCESS

Many corporations clearly see the need for and understand the opportunities of collaborating with data scientists. Creating a space for playing with data, whether by supporting a hackathon or offering free access to technology that is normally proprietary, should be encouraged. Companies benefit from promoting their brand, obtaining street credibility for tackling problems that matter to society, and gaining a view of the latest trends in big data.

A related revolution is taking place in open-access modeling, whereby organizations are able to share access to certain modeling capability for free. For varying reasons, organizations recognize the advantages of collaborating and sharing resources. The insurance industry has initiated an open access catastrophe-modeling platform that allows researchers and practitioners to plug and play different modules. Just as model performance can be improved by using a multitude of approaches, it is also possible to improve the mechanism for collecting data. Crowd-sourcing is now a popular method for collecting data about a variety of issues, from risk of natural disasters to medical disorders to information about the universe. Furthermore, a number of scientific experiments have been possible only because of sufficient public interest and a willingness to help researchers by using screensavers to run models on millions of personal computers. For example, climateprediction. net is one such volunteer computing project that helps answer questions about how climate change is affecting the world.

ENSURING PERSONAL PROTECTION

The cloud is is intertwined with big data. Rather than worry about managing large volumes of information, organizations are becoming increasingly reliant on cloud-based solutions for both storing and processing data. Dropbox, Box and Google Drive offer storage solutions that facilitate sharing of data. Amazon Elastic Compute Cloud (Amazon EC2) provides a scalable approach to cloud computing with flexible on-demand billing. While outsourcing the challenge of collecting, storing, processing, and utilizing data to those organizations that are seen as experts, many fear that this trend is creating large security risks.

The big question at present concerns whether people will pay for privacy in the light of increasing commercial use of personal data and increasing numbers of security breaches. There are various possible approaches that might be used for making computer systems more secure, but most come with significant costs. One such approach, known as *homomorphic encryption,* allows computational processing to be carried out directly on encrypted data, but the processing speeds are currently much slower than on unencrypted data.

PRIVATE CLOUDS

Many organizations want to have the benefits of cloud computing but with greater levels of security and control. A *private cloud* is a secure cloud-based environment protected by a corporate firewall. This approach is similar to the more traditional model of individual local access networks (LANs) but with the added advantages of virtualization, making it possible to access multiple versions of servers, operating systems, storage devices, and other IT services.

SANITIZING DATA

Even databases that have been anonymized are vulnerable to what are known as "linkage attacks." By meticulously comparing data from a private database with details scraped from open sources, such as social media websites, it may be possible to match entries such as dates of birth, addresses, nationalities, and occupations. *Differential privacy* refers to the objective of optimizing the usability of sensitive databases while minimizing privacy risks. One promising approach is to use mathematical algorithms to add noise to the data in order to disguise the identity of individuals. At present, there is no agreed standard for differential privacy, but a reliable solution is necessary to encourage more organizations to provide open access to data.

EVIDENCE-BASED POLICY

Policy design is a serious challenge to all governments and, when done correctly, can deliver substantial benefits for those countries. Few policy-makers would admit that their policies were based on a whim or a gut feeling, and of course there is usually good motivation for rolling out new policies. On the other hand, history has recorded many cases where policies were rejected and eventually failed. Of course, it makes sense to test new policies before subjecting an entire nation to change. However, no region or group of citizens wants to be the guinea pigs within the government's experimental laboratory.

The U.K. conservative government's Community Charge, better known as the Poll Tax, trial in Scotland in 1989 serves to illustrate what can happen when policies are not accepted by a large section of society. In this case, the prime minister, Margaret Thatcher, piloted an extremely unpopular and untested policy that would impose a flat tax on every adult. She imposed this tax on a part of the country where she had the least to lose, given the low level of support for her government in Scotland. This experiment led to social chaos in the form of riots, strikes, and general disorder, which disrupted the country for months and eventually led to the downfall of the prime minister herself.

Policymakers are faced with the task of making important decisions in the present that will determine how we cope with both known and unanticipated challenges over the next few decades. More multidisciplinary research and cooperation between scientists, the private sector, and government is needed. While decision-makers would prefer to obtain a simple scenario, it is important to accept that the future is inherently uncertain and no crystal ball exists. A thorough evaluation of all the quantitative and qualitative information is likely to produce numerous future scenarios.

Big data offers new intriguing possibilities that go beyond evidence-based policymaking. As social media starts to provide a mechanism for harvesting opinions and ideas, a potential approach is to use data and quantitative modeling to construct, evaluate, and fine-tune policies. The U.K. government's e-petition website collects data about the issues that are of most importance and has the potential to form the basis of a big-data laboratory. Taha Yasseri, Scott Hale, and Helen Margetts ("Modeling the rise in internet-based petitions," Oxford Internet Institute, 2014) study hourly data from e-petitions and demonstrate the importance of the first day in determining success.

An interesting example of the use of big data generated by the Internet and social media for influencing policy is the presence of citizens' lobbying groups, such as 38 Degrees in the U.K. and MoveOn in the U.S. These websites provide online campaigns and the power of mass media to rally support or protest around key policies. In the U.K., 38 Degrees successfully led the government to change its policies in key areas such as the privatization of forestry, and MoveOn is credited with helping Barack Obama become the first African-American president of the United States.

PUBLIC-PRIVATE PARTNERSHIPS

As data revolutionizes the way governments make decisions and inform policy, many developing countries are held back due to a lack of infrastructure for collecting, processing, and utilizing data and by a shortfall of experts. When attempting to assess the impacts of global challenges such as climate change, their governments are at a serious disadvantage. This is particularly problematic when quantifying the risk of extreme events, for which historical records of a long duration are required. There is a growing need to bring together researchers and practitioners with expertise in environmental science, statistics, physics, mathematics, engineering, and economics to develop a multidisciplinary approach for quantifying risks; analyzing the impacts, costs, and benefits of competing policies; and establishing strategies for increasing resilience. Public-private partnerships involving cooperation between government and industry are demonstrating that collaboration across many sectors of society will be necessary to provide adequate solutions and inform policymakers.

The Alliance of Small Island States (AOSIS) is an intergovernmental organization that is working to address the challenges of global warming, focusing
primarily on the threat of sea level rise, by representing the interests of
low-lying coastal and small island countries throughout UN negotiations on
climate change. Its members have found that the lack of historical data for
their countries places them at a disadvantage when negotiating with developed countries. In order to improve understanding of the impacts of climate
change, the international community is working, through organizations such
as the World Bank and the United Nations International Strategy for Disaster
Reduction (UNISDR), to promote risk assessment projects and encourage the
integration of insurance into disaster risk-management strategies. While the
reinsurance industry has considerable experience using historical data and
catastrophe modeling techniques for quantifying and pricing risk, this
knowledge is rarely available for developing countries.

Fortunately, a number of public-private partnerships are working to change
this situation:

- In response to the lack of open-access catastrophe models, the Organisation for Economic Co-operation and Development (OECD) started
  the Global Earthquake Model (GEM, www.globalquakemodel.
  org) in 2006. This €35-million project aims to measure and communicate earthquake risk worldwide. GEM considers the hazard, exposure,
  vulnerability, and socioeconomic impact of earthquakes.
- The Willis Research Network (WRN, www.willisresearchnetwork.
  com) supports open academic research and the development of new risk
  models and applications. It aims to provide an open forum for promoting
  scientific research on extreme events, primarily climate and weather risks,
  through collaboration between universities, insurers, reinsurers, catastrophe modeling companies, government research institutions, and non-
  governmental organizations.
- The Lighthill Risk Network (www.lighthillrisknetwork.org)
  promotes collaboration between insurers, reinsurers, and brokers and
  facilitates knowledge transfer into business from academic, government, professional, and commercial experts at the forefront of risk-
  related research.
- Collaboration across the insurance industry, government, and academia
  has led to the formation of the OASIS Loss Modelling Framework
  (www.oasislmf.org), which has the potential to deliver a larger
  choice of models, a better understanding of their performance, and the
  ability to combine different modules from a variety of vendors and
  academic researchers which will encourage greater cooperation.

The search for solutions for adapting to climate change and building
resilience in the face of increasing climate variability has led policymakers

to actively engage with scientists and the private sector. Along with the innovative public-private partnerships described earlier, new insurance products have been created. *Parametric insurance* is an alternative type of insurance that agrees to make a pre-specified payment immediately after a triggering event occurs. The triggering event could be a natural catastrophe or a situation where an index, such as a weather variable, exceeds a specific threshold and is expected to lead to financial losses. Parametric insurance products provide a means for using information based on independent sources of data and offers a mechanism for government and the private sector to form partnerships. Catastrophe programs are examples of where public-private partnerships have been created to leverage the use of parametric insurance.

Catastrophe programs are particularly relevant for transferring risk in developing countries, as the international community often plays the role of reinsurer of last resort in the aftermath of a natural disaster. The need for adequate disaster risk financing in developing countries has received growing attention due to the major economic risks arising from climate change (Eugene Gurenko, *Climate change and insurance: disaster risk financing in developing countries,* Earthscan, 2007). Many of these risk financing programs are supported by donors and international financial institutions. Substantial challenges still exist to achieving competitive catastrophe risk markets in developing countries, requiring assistance from the insurance industry (J. David Cummins & Olivier Mahul, *Catastrophe risk financing in developing countries,* World Bank, 2009). The Caribbean Catastrophe Risk Insurance Facility (CCRIF, www.ccrif.org) is the first multi-country risk pool based on parametric policies backed by both traditional and capital markets. While many of these catastrophe programs were created as public-private partnerships, the specific approaches are diverse and range from no state involvement to unlimited state guarantee and may be compulsory or voluntary.

In practice, many programs have been initiated in response to major catastrophes rather than a planned approach to manage disaster risk. For example, the Florida Hurricane Catastrophe Fund (FHCF) was initiated in Florida after Hurricane Andrew in 1992, the Turkish Catastrophe Insurance Pool (TCIP) was created in response to the Marmara earthquake of 1999, and the California Earthquake Authority (CEA) was developed after the Northridge earthquake of 1994. These responses support the argument that disaster risk management and the consideration of a catastrophe program as a means of ex-ante policy rarely becomes a priority until the aftermath of an extreme catastrophic event. Unfortunately, as a result, the majority of these programs have been established at a time when insurance and reinsurance costs are likely to have increased.

Fluctuations in weather patterns could exacerbate poverty by undermining the livelihoods of the poorest people in developing nations who depend on climate-sensitive sectors such as agriculture, forestry, or fishing for their income. For many of these people, insurance is not accessible, whether due to cost or lack of adequate systems. Parametric insurance is particularly attractive for managing environmental risk and for providing protection in the agricultural sector. It has been successfully piloted for both development and disaster relief. Ethiopian farmers have the option of working on irrigation and forestry projects or using cash to pay for insurance products, developed by the UN's World Food Programme (WFP) and Oxfam America and reinsured by Swiss Re, which will compensate the farmers if a severe drought affects their crops. Index-based insurance initiatives are removing the barriers of climate risk at the level of individuals, banks, cooperatives, and government in many countries, including India, Mongolia, China, Nicaragua, and Thailand. Alongside other risk management policies, index-based insurance offers a means of building adaptive capacity and resilience through risk transfer, increasing access to credit and incentivizing risk reduction via appropriate price signals.

Index-based insurance faces many challenges, such as the need for training and education about this relatively new form of insurance and the behavioral change required for its acceptance. In practice, the accuracy of the index is constrained by the duration and quality of the available data, which is a considerable challenge in developing countries. Inadequate modeling and a lack of data may result in what is known as basis risk, which refers to the fact that losses may be sustained while the insurance product does not provide adequate compensation.

There is a role for governments and aid agencies to fund the infrastructure required for collecting meteorological, loss, and socio-economic data in order to evaluate the feasibility of index-based insurance. Having this infrastructure in place could provide ample opportunity for the private sector to create new parametric insurance products and is also of great importance for assessing climate variability and long-term environmental change. In practice, index-based insurance will be truly successful only if its pioneers manage to strike a balance between the complexity of the model required for constructing the index and the simplicity and transparency needed to effectively communicate its benefits to the producers. For example, further research is required to evaluate the feasibility of using satellite data and weather sensors to measure the impact of weather events on agricultural production.

IMPACT BONDS

Imagine going to a cash-strapped government offering to raise funds to finance an innovative social or development project with tangible delivera-bles linked to transparent key performance indicators. This is exactly the point of impact investing, and the selling point for investors is that they receive a decent rate of return for taking the risk of financing a project that aims to deliver social good. The ability to both enable a positive change in the world and make money at the same time is proving very attractive, especially at a time when interest rates are so low.

SOCIAL IMPACT BOND

A Social Impact Bond (SIB) is a specific type of social impact financing in which funds are raised from investors to provide social service provider(s) with the working capital to deliver their services. In short, a SIB is a form of public-private partnership based on a payment-by-results contract where investors provide funds for a not-for-profit organization to carry out interventions. SIBs, also referred to as social innovation financing, known as Pay for Success Bonds in the U.S. and Social Benefit Bonds in Australia, bring the rigor and attention to detail of the financial sector that can help make these complex social projects a success. By monitoring KPIs in real-time, the project managers can dedicate resources to areas where more attention is required and, of course, the incentives to succeed are sufficiently high. SIBs help to establish a financing incentive for providers to focus on outcomes, rather than on process and reporting. By unlocking private investment in social outcomes, SIBs have received a lot of attention and are viewed by some as the future of funding social change. The government or other sponsors pay out if the interventions are successful. The investors make a profit if the project is a success as measured by specific KPIs and lose money if it does not.

The Peterborough SIB was launched by Social Finance UK in 2010 and was the first financial investment that aligned successful social outcomes with financial returns. Having raised £5 million from social investment, it is designed to reduce reoffending among short-sentence male prisoners. Investors stand to receive a rate of return of up to 13 percent with an eight-year investment horizon, depending on the success of the intervention. The Peterborough project works with three cohorts of 1,000 offenders and aims to reduce the level of offending by at least 7.5 percent compared to results in a national control group. The Big Lottery Fund and the Ministry of Justice are providing the payments investors receive if the project is successful.

The SIB model is widely viewed as a success, with 14 such programs currently underway in the U.K. There are now 14 SIBs in the U.K., five in the U.S., two in Australia, one in the Netherlands, and one in Belgium, and more than 100 proposals worldwide. Over $100 million has been raised globally in social investment to fund SIBs covering a wide range of issues such as housing, recidivism, employment, and educational outcomes. For example, Goldman Sachs invested approximately $10 million to help fund a SIB in New York City in partnership with the City of New York, Bloomberg

Philanthropies, and MDRC (a social services provider). This SIB, called Adolescent Behavioral Learning Experience (ABLE), is aimed at reducing the recidivism rate for adolescent offenders at the Rikers Island correctional facility.

Early figures published in 2013 suggest that the Peterborough SIB has led to an 11-percent reduction in reoffending over the period of the pilot, against a 10-percent increase nationally. Furthermore, organizations participating in the project have reported greater innovation in their methods for tracking and improving the performance of their services. This evidence suggests that applying big data analytics to projects that provide societal benefits could be the basis of a variety of innovative financial instruments.

The future for Social Impact Bonds looks bright, as the U.K. Cabinet Office launched a new £30-million package in June 2014 to back SIBs to help disadvantaged young people get into education, work, or training. This package includes creation of a Social Impact Bond Centre of Excellence to support the development of more SIBs across the public sector. The private sector is also taking notice of this opportunity. Bridges Ventures runs a £14-million SIB investment fund, and the SIB design specialist Social Finance has steadily grown its practice.

DEVELOPMENT IMPACT BONDS

Much larger investment opportunities could be generated by designing impact bonds for international development, and this potential was recently explored in 2013 by Social Finance UK and the Center for Global Development. They concluded that there is a considerable opportunity to "transform social problems into 'investible' opportunities and create incentives for investors to put in place the necessary feedback loops, data collection and performance-management systems required to achieve desired outcomes."

A Development Impact Bond (DIB) is a variation of the SIB that provides new sources of financing to achieve improved social outcomes in developing country contexts. As with SIBs, investors provide external financing and receive a return only if pre-agreed outcomes are achieved. Funds to remunerate investors come from donors, the budget of the host government, or a combination of the two. Financial returns to investors are intended to be commensurate with the level of success. DIBs have the potential to improve aid efficiency, cost-effectiveness, and accountability by shifting the focus onto implementation quality and the delivery of successful results. By creating partnerships between emerging nations, aid donors, and private investors, it is hoped that development spending can have the greatest possible value for money.

The U.K. government's Department for International Development (DFID) is preparing to launch its first ever DIB to improve healthcare in Africa by bringing together private and public investment. The new DIB will invest in the prevention of sleeping sickness in Uganda, and DFID will spend £1.5 million on research and design of the DIB. Its specific objectives are to significantly reduce the risk of sleeping sickness and provide an incentive to innovate area livestock treatment to manage the prevalence of sleeping sickness in people. According to the U.K. government website, DFID will also "launch a new roadmap for increasing further private investment into frontier developing economies." In order to share the findings of this revolutionary approach with the international development community, DFID will also set up a new online open-source knowledge platform and help to convene new development cooperation hubs in Nigeria, Mozambique, and Kenya, where private companies, governments, international organizations, and civil society organizations are able to collaborate on joint development projects.

DIBs could have substantial potential for programs in global health, education, and job creation where results can be measured within several years. The first DIB in education was launched in 2014 by the Children's Investment Fund Foundation to address the fact that 40 percent of girls in India do not remain in school past the fourth grade. Managed by Instiglio, this DIB has the aim of increasing enrollment of girls and improving learning outcomes for all children in Rajasthan, India. The Mozambique Ministry of Health in partnership with Anglo American, Nando's, and Dalberg Global Development Advisors have launched a DIB and plan to raise up to US$700 million to fund 12 years of malaria interventions reaching up to 8 million people in order to reduce malaria prevalence in targeted areas by up to 75 percent. Investors will obtain a 5-percent return if incidence rates fall by at least 30 percent by year three, and are repaid half of their principal if the program is unsuccessful.

THE ROLE OF BIG DATA

There is a crucial role for big data in helping to structure impact bonds and manage the expectation around what can be feasibly achieved. Big data is at the heart of the process for facilitating the design, implementation, monitoring, and evaluation of these bonds. As the timeframe for impact bonds, both SIBs and DIBs, is generally over several years, it is difficult to say how successful they will be in efficiently delivering outcomes in the long run. There is no doubt that the storyline is attractive and that early success for national SIBs heralds good future prospects. But lessons could be learned from the implementation of index-based insurance projects where, in many cases, little attention was given to the important details of ensuring that the index was based on relevant data and that a quantitative model could be constructed to map between this index and the actual losses. As a result, many index-based insurance products have failed to be sustainable after the original donors' funds were removed.

Without the input of data scientists and use of appropriate analytics, it is very likely that some impact bonds will fail due to design faults. It is critical to have access to high-quality data that has been independently validated and to develop an accurate quantitative model to understand what is achievable when structuring these bonds. If either the data or model is inaccurate, the project could fail to deliver adequate social impact, and these bonds will then cease to attract investors in the future. Irrespective of the desire to make the world a better place, it takes only one case of a poor rate of return to spook investors.

EGYPT, 2011 (CONTINUED)

Many reasons could be listed to attempt to explain the cause of the Arab Spring. While dissatisfaction arising from being ruled by dictators, autocrats, and absolute monarchs is a prominent factor, a variety of other issues such as poverty, income inequality, political corruption, unemployment, economic recession, and human rights abuses are commonly cited. An appealing explanation that aligns with the high participation of young people and use of social media in organizing events is the demographic factor that results from having a large percentage of educated but dissatisfied youth. As data is generated from the interactions between government and its citizens, we can aspire to use this information to improve governance structures and create a democratic knowledge-based society where the voices of everyone can be heard.

8 CORPORATE SUSTAINABILITY

CITY OF LONDON

COVERING JUST ONE square mile and providing employment for 400 thousand people, the City of London is a leading global financial center. A total of 588 foreign companies are listed in London, representing 20 percent of the global foreign equity listings. According to the 2014 "Cities of Opportunity" report conducted by multinational professional services network, PwC, London was named as the best city overall and the only city to finish first in three out of the ten indicators: economic clout, technology readiness, and city gateway. These indicators are based on publicly available data from the World Bank, the International Monetary Fund and other national statistics organizations. While cities compete for the highest rankings on different indicators, big data is also providing a detailed perspective of the activities of corporations and helping to quantify sustainability.

One of the major hallmarks of the 21st century is the level of interconnectivity that now exists due to the ease of travel, virtually free voice and video communication, and global trade networks. There is no doubt that innovations brought about by the Internet, mobile telephone, and social media have made the planet appear smaller than ever before. While we all enjoy instantaneous contact with family, friends, and colleagues from wherever we happen to be situated in the world, it is important to ask what effect, if any, these changes are having on the private sector.

Sustainability can be guaranteed only if we, as a global community, figure out how to utilize and share our limited resources more efficiently. Security of supply for key resources such as water, food, and energy is fortunately receiving increasing attention. The traditional linear take-make-dispose model is being abandoned in favor of a circular economy where goods are recycled and repurposed in order to reduce waste and generate potential for high added value. The concept of natural capital helps to emphasize that a sustainable world economy can be achieved only if we start to value natural resources and the ecosystem services that we rely upon.

Data will play an important role in helping to measure the environmental impacts and quantify environmental risks associated with different decisions, actions, and operations within business, supply chains, and the international community. The United Nations advocates measurement, reporting, and verification (MRV) as a means of monitoring nations' attempts at addressing the challenges of climate change. While reducing carbon emissions is of foremost importance for the international community in trying to halt global warming, individual nations are working hard to promote sustainable development.

The corporate world is also beginning to pull its weight. Despite accusations of green-washing and managers talking about sustainability simply as a means of PR without much substance, there is growing evidence of organizations realizing that management practices must change if companies are to be resilient to climate change and also have a chance of identifying potential opportunities. The introduction of indicators for measuring environmental, social, and governance (ESG) factors offer considerable potential for change since companies scoring high on ESG indicators are believed to have superior business models. Data provides a means of creating dashboards for monitoring and comparing ESG factors and other measures of sustainability. Without access to such dashboards and the strategic guidance that they offer, it will be difficult to encourage behavior change within corporations.

GLOBAL MEGAFORCES

C-level executives of large corporations have many pressing issues to worry about. Due to the pressures arising from demanding shareholders, many of the concerns that are considered a priority tend to relate to short-term performance. After all, who has time to worry about the next 20 years if the firm might fail in the next year? In contrast, the planet and its societies are already facing environmental challenges that require long-term planning and decision-making. As information about risks and opportunities becomes available and the uncertainty slowly decreases, it is likely that sufficient funds will be invested now to prevent future adverse impacts from being realized.

While academic researchers publish and disseminate the best available scientific knowledge, management consultants are required to advise on how strategies based on this scientific knowledge can be formulated and opera-tionalized by their clients from a range of different sectors. KPMG International has identified ten megaforces that will significantly affect corporate growth globally over the next two decades. Their 2012 study "Expect the Unexpected: Building business value in a changing world" explores issues such as climate change, energy and fuel volatility, water scarcity, and exploitation of natural resources, as well as population growth spawning new urban centers. The analysis examines how these global forces may impact business and industry, calculates the environmental costs to business, and calls for business and policymakers in government to work more closely together to mitigate future business risk and act on opportunities.

In what follows, we focus on three of these megaforces (population, carbon emissions, and water) and the challenges of obtaining sufficiently high-quality data to facilitate evidence-based decision-making. As big data becomes available to track and monitor the impacts of these megaforces as well as the business activities that gives rise to them, business managers will be equipped with the tools and techniques required to establish appropriate strategies and motivate behavior change.

POPULATION

It is only fair to start with the source of the problem, which is the sheer amount of people who are using and depend on the planet's resources. Having risen steadily from 350 million at the time of the Black Death (one of the most devastating pandemics in human history killing around 20 million people) in 1350 to 7 billion today, the UN projections suggest that the world population could exceed 10 billion by 2100. A planet with more people implies greater demand for water, food, and energy. As the size of the middle class grows in developing countries, the demand for technology will place further strains on energy supplies.

Some of the mismatch between demand and supply can be solved by greater efficiency and waste reduction, and there is a substantial role for using data to transform business operations. In developed countries, there are great expectations that technology will get us out of this fix without having to drastically reduce consumption and give up some of our creature comforts.

CARBON FOOTPRINT

Greenhouse gas emissions are now known to be responsible for global warming caused by humans. These greenhouse gases include carbon dioxide and methane, and it is common practice to calculate a carbon dioxide equivalent. *Carbon footprint* refers to the total amount of greenhouse gas emissions caused by an organization, event, product, or person.

Obviously, the ratio of the profits generated by a firm to its carbon footprint is a crucial indicator for investors. Oil and gas exploration and other organizations that use fossil fuels are an important component of many financial portfolios and indeed have been responsible for stellar track records over the years. It is difficult to reconcile a desire to persuade organizations to stop polluting with the expectation that pensions will perform well enough to support an aging population in the developed world. With the help of big data analytics, the outcomes of these decisions will be clearer.

WATER SCARCITY

Benjamin Franklin wrote, "When the well's dry, we know the worth of water." Water is essential for life, and clean water is critical for ensuring good health. Although the surface of the planet is two-thirds water and only one-third land, access to sufficient quantities of water for agriculture, industry, and human consumption is already posing considerable challenges. The UN estimates that almost one-fifth (1.2 billion people) of the world's population live in areas of physical scarcity and another 500 million people are approaching this situation. The planet has sufficient freshwater for 7 billion people, but it is unevenly distributed and much of it is wasted, polluted, and unsustainably managed.

Companies that rely on large quantities of water, such as the food and beverage sector, are already facing challenges with regard to water scarcity. It is now a priority for the many firms that are addressing ESG issues to make sure that the environment is protected, local communities are fairly compensated, and business practices are sustainable in the long-run. Interestingly, while carbon would seem like the most obvious focus for those interested in tackling climate change, it has not had the impact on business that was initially expected, due to the failure of attempts by the international community and the global market to deliver a clear price for carbon. In contrast, water scarcity has been a more direct catalyst for action and is receiving increasing levels of attention.

ENVIRONMENTAL RISK

BP AND EXXON MOBILE

On March 24, 1989, the *Exxon Valdez* oil tanker struck Prince William Sound's Bligh Reef and spilled between 260,000 and 750,000 barrels of crude oil, making it one of the most devastating human-caused environmental disasters. The *Valdez* spill was the largest ever in U.S. waters until the 2010 Deepwater Horizon oil spill, in terms of the volume released. Exxon had sufficient time to manage the cleanup and deal with the environmentalists who were primarily concerned with the disaster.

On April 20, 2010, the explosion of Deepwater Horizon, which drilled on the British Petroleum operated Macondo Prospect, killed 11 men and injured 17 others. The resulting oil spill continued unabated for three months in 2010 and represents the largest accidental marine oil spill in the history of the petroleum industry. On July 15, 2010, the leak was finally stopped by capping the gushing wellhead after it had released about 4.9 million barrels of crude oil.

The resulting impact of these otherwise comparable disasters on the two organizations (Exxon and BP) is striking and shows how environmental risk has become an increasingly significant threat to the resilience of organizations. Global communications via the Internet and social media ensured that what started as an environmental disaster rapidly became a source of reputational risk for BP. The disaster led to the resignation of its chief executive, Tony Hayward. BP's total provision for the disaster is $42 billion but the cost could reach $90 billion if the maximum possible penalties and damages are awarded.

The signature of this new environmental risk is best seen in the dramatic differences in the response of the companies' share prices to these two disasters. A quantitative comparison of the impacts of these two oil spills on share prices for BP in 2010 and Exxon in 1989 demonstrates that the financial world now places a substantial risk premium on environmental disasters of this kind. By anchoring the share prices of Exxon and BP at 100 on the last day of trading before the respective disasters, it is possible to investigate the impact of the disasters and the consequent events on their shareholders (as shown in the preceding figure). While Exxon's share price dropped by 6 percent, that of BP declined by 53 percent before starting to recover. This suggests that risk of a share price crash due to oil spill catastrophes has increased by a factor of nine in the last two decades. In terms of recovery rates, Exxon's share price returned to the same level as before the spill after only 20 business days. In contrast, BP's share price is still suffering from the distaster (over four and a half years later) and currently sustains a loss of 30 percent.

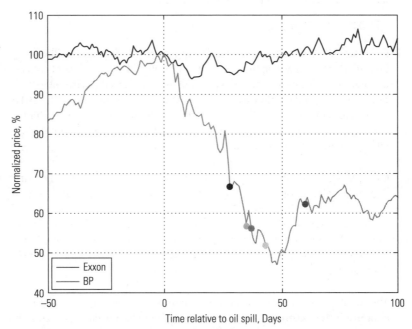

Normalized prices for Exxon and BP on trading days before and after the oil spills of Exxon Valdez (3/24/1989) and Deepwater Horizon (4/20/2010). From left to right, the dots on the BP prices indicate the impact of specific events: "top kill" operation was unsuccessful; U.S. government signals it will take legal action to force BP to stop paying a dividend to shareholders; U.S. President Obama compares the oil spill's impact to that of the 9/11 terror attacks; BP CEO Tony Hayward hands day-to-day control of the spill operation to Bob Dudley; and BP announces that it has stopped its Gulf of Mexico leak for the first time since April.

EARLY WARNING SYSTEMS

Big data will provide sufficient information for constructing mathematical models of a variety of critical complex systems. Ranging from patients in intensive care units to nuclear reactors to the planet's ecosystem, access to data from multiple sources enables switching from being a passive observer to monitoring risk in real-time, and eventually developing a feedback loop that can stabilize, control, and better manage such complex systems. An important initial component of this process of enlightenment through big data is the development of early warning systems.

Many of us have witnessed the use of "traffic light" warning systems (green, yellow, red) indicating the level of warning in different critical situations such as extreme weather alerts or terrorism threats. Each traffic light represents a measurement of the probability of an extreme event occurring in the near future. The traffic lights are obviously a simplified representation of the probability, but they are enormously helpful in coordinating an appropriate response to the threat. Furthermore, the process involved in measuring the probability is often qualitative, based primarily on the occurrence of specific events. For this reason, the traffic light system helps to distill and communicate information about the situation and whether or not action should be taken.

As more and more sources of data become available, there is substantial potential to use patterns of past adverse events and, to develop a quantitative model that can be used to assess the risk level. This approach also has the advantage of considering different response actions and their associated economic costs and benefits. Taking such a quantitative approach allows for constructing appropriate boundaries for the probability levels and the optimal number of colors in the traffic light.

EDF Energy owns a nuclear reactor at Torness in Scotland, which is situated on the coast and uses a water cooling system. Unfortunately, the reactor has been shut down many times due to clogging events caused by marine species such as seaweed and jellyfish being sucked into the cooling system. The cost of an unexpected shut-down is several times greater than an orderly shut-down. Empirical evidence for increasing numbers of jellyfish likely caused by global warming suggests that this problem will probably intensify in the coming years. EDF Energy has had the foresight to investigate the development of an early warning system based on environmental observations, weather forecasts, and operational data relating to the water-filtering process. This is a multidisciplinary undertaking involving not only data from different sources, but also mechanistic models about the natural movements of jellyfish and seaweed.

SOCIAL MEDIA

Before the Internet, it was possible for remote isolated events to stay completely unnoticed or to take a long time before hitting mainstream media. This led some corporations to ignore the adverse impacts that their business operations might be having on the environment and livelihoods of local communities. Without media attention, these externalities were often negelected and not seen as being relevant to building a sustainable business.

If they were not affecting the balance sheet, then surely stakeholders were happy and everything was okay.

Social media brought global interconnectivity, and with it came a global conscience. Slowly, consumers started to understand how their choices in the developed world were impacting the lives of others in the developing nations. For example, the ability to purchase cheap clothes was linked to poor working conditions and child labor in countries like Bangladesh. When the Rana Plaza factory building near Dhaka collapsed in 2013 and killed over 1,000 people, it was clear that the purchase of cheap clothes was having a direct effect on the risks imposed on these employees.

The horsemeat scandal in the U.K. is another example of where a news story quickly spread and caught retailers off their guard. Consumers discovered that certain meat products, labeled as containing beef, actually contained horse. Although this mislabeling was not a danger to health, as the citizens of several countries happily eat horse meat, there was public outrage as it became apparent that major retailers did not understand or adequately scrutinize the quality of their supply chains.

A new category of risk, known as *reputational risk,* is now of particular interest for corporations. When news broke that large multinational corporations such as Starbucks, Amazon, and Google were dodging taxes in 2013, social media was used to organize campaigns to lobby for change. Despite these corporations insisting that their tax affairs were legal, it quickly became apparent that the exploitation of tax loopholes was seen as immoral by some customers and was resulting in loss of business due to protests. Starbucks quickly responded to this growing reputational risk by agreeing to pay £5 million ($3.7 million) worth of U.K. corporation tax, the first payment made in five years.

Rather than hoping that such issues do not arise again, some companies are now actively seeking out potential risks with the intention of intervening before they become unmanageable. One way to achieve this proactive response using data is the automated surveillance of social media, blogs, and newsgroups around the world. Start-ups such as RepRisk provide such a service that acts as an early warning system for managers. RepRisk helps clients proactively assess ESG concerns that could potentially cause financial, reputation and compliance issues.

RISK AND RESILIENCE

Approaches to quantifying and assessing risk form the bread and butter of insurance organizations. Risk reduction strategies assume that the sources of risk can be identified, assessed, and managed. Whether through risk mitigation or adaptation, it is expected that these risks can be controlled. Resilience, on the other hand, presents a more holistic approach. The word itself is certainly more optimistic — better to talk about building resilience than managing risk, which acknowledges weakness. A resilient entity, be it an individual, organization, city, or nation, can be defined as one with an ability to absorb, react, or transform when faced with adversity. Adaptation and mitigation are approaches for absorbing and reacting to threats. Transformation may be an uncomfortable outcome, as it is likely to be expensive and certainly disruptive as it acknowledges the fact that absorbing and reacting are no longer sufficient mechanisms. Use of big data and adequate modeling will allow organizations to determine strategies for building resilience.

Big data is already generating disruptive change, leading to the extinction of some traditional business models and opportunities for others. In business, there are many examples of firms that did not manage to transform in time. Kodak failed to recognize and manage the considerable threat posed by digital photography. Similarly, there are examples of organizations that had the foresight to transform, embrace change, and grasp opportunities before their competitors: IBM went from selling computer hardware to become a global consulting firm; the mobile telephone company Nokia started life as a Finnish paper mill, rubber works, and cable works company; GE was originally a vehicle for Thomas Edison to sell his electric inventions; Apple extended its business from producing Mac computers to creating iPads and iPods and delivering online music through iTunes.

MEASURING SUSTAINABILITY

As social media ignites a global conscience, an obvious question for managers to ask themselves is, "How is my business doing?" Share prices and quarterly accounting figures provide sufficient financial metrics, but what about more intangible aspects such as environmental, social, and governance (ESG) principles?

In the 2010 series of interviews and surveys conducted by Accenture and the UN, 70 percent of CEOs reported that they are integrating ESG issues into their core business strategies "more" or "much more" than five years ago. It is likely that the temptation for "green-washing," exaggerated marketing of environmental policies, may have been driving some of the responses. To be fair to the CEOs, it is not easy to navigate the sea of confusion surrounding sustainability; as to date, there is no recognized standard for monitoring or evaluating the ESG principles of a particular company. There are clearly leading lights such as Unilever, PUMA, and Marks & Spencer, which have worked hard at promoting models for sustainable business practices. In order to scale up this activity, managers require both guidance and strategies, and investors need access to key performance indicators for ranking the achievements of competing firms.

Luckily, a number of initiatives and start-ups are now collecting relevant information about businesses and their operations with the aim of promoting sustainability: The Carbon Disclosure Project (CDP), Trucost, Sustainalytics, and Carbon Analytics, CDP, an investor engagement initiative, aims to encourage organizations, cities, and governments to measure, disclose, manage, and share vital information about impacts on the environment and natural resources. By promoting awareness through self-reporting, it is hoped that this will drive decision-making to lower emissions and increase sustainability.

Trucost aims to help its clients understand the economic consequences of an unsustainable dependence on natural capital. It does this by identifying environmental risk and opportunity across company operations, supply chains, and investment portfolios. Sustainalytics is a global responsible-investment research firm that helps investors integrate ESG information and assessments into their investment decisions.

Carbon Analytics is helping organizations to quantify and visualize their carbon footprint and emissions at a range of levels, from the company itself right down to its supply chain. Using big data and sophisticated modeling techniques, it is expected that this innovative approach will be a game changer for many businesses that want to develop strategies about the optimal path for reducing carbon emissions.

LONG-TERM DECISION MAKING

It is only too easy to blame the investors for being cold-hearted when it comes to having concerns about the planet, ecosystem, and livelihoods of communities living in developing countries. It is likely that behavior change may be brought about within the investment community only by demonstrating a direct link with future profits of individual firms rather than expecting investors to be able to start making decisions that benefit society at large.

According to the NYSE Factbook, over the last few decades, the average holding period for equities has steadily decreased from eight years in 1960 to a little more than a year in 2000 to less than four months in 2014. While this is likely due to an increase in day trading and high-frequency automated trading, it suggests that many of these shareholders are not concerned about the long-term value of their investments. Furthermore, many institutional investors are constrained by fiduciary responsibilities to utilize funds that track the market, which is currently dominated by firms that are both carbon-intensive and natural resource-intensive.

Many sections of society are now actively voicing their concerns about climate change and the need for immediate action to avoid global warming of two degrees Celsius or more. This has led to divestment campaigns focused on calling for institutional investors to cease investing in organizations involved in the exploration or extraction of fossil fuels such as oil and gas. As this movement gains traction, financial institutions are struggling with the best way to adapt.

As climate change delivers challenges over the next few decades, the financing of appropriate mitigation and adaptation solutions relies on foresight and associated long-term investments. Fortunately, there is one group of investors that is taking a lead. Pension funds typically hold positions for an average of eight years and therefore give serious consideration to both rewards and risks over long time frames. As scientific consensus builds around the damaging consequences of climate change, pension funds are advocating for sustainable investment.

There is now solid evidence that these developments are having an effect. For example, the Norwegian pension fund and insurance firm, Storebrand, with assets under management worth $80 billion, have divested from 23 fossil fuel companies, saying in the future these stocks would be "worthless financially." In addition, some green funds have a mandate to outperform an appropriate equity index, but without holding certain stocks that have been screened due to poor ESG. Rather than seeing such screening as a handicap, the fund managers view the ESG information as a means of improving long-term performance. Big data about ESG issues will be crucially important in facilitating this move towards responsible investment.

STRANDED ASSETS

One might expect discussions around opportunities rather than risks to sway investors, but at present, it appears that fear rather than greed is causing a change of heart. The accounting term *stranded asset* refers to an asset that has become obsolete or non-performing, resulting in an unexpected devaluation or conversion to a liability. The risk of asset stranding is increasingly related to environmental risks such as regulation, clean technology, resource constraints, and potential litigation. While many fund managers admit to wanting to address the challenges of climate change through their investment decisions, the fact is that their hands are tied due to investment policies. However, measuring the risks of assets becoming stranded due to a range of future scenarios could level the playing field and make certain equities seem less attractive as long-term investments than before.

As big data about ESG metrics and information about the particular projects that create value for individual firms becomes available, it is likely that long-term investors such as pension fund managers will become increasingly discerning with their decisions and investments. If such a shift takes place in a sufficiently large number of investment funds, it could signal a change that is likely to flow over to many other institutions within the financial sector. Indeed paradigm shifts such as this have always had slow beginnings, as has been the case for the anti-slavery and anti-smoking movements.

CITY OF LONDON (CONTINUED)

Data will play a prominent role in measuring environmental impacts and risks associated with different business strategies. In a socially-connected world, an incorrect decision can quickly cascade ramifications across a broad population, doing irreparable harm to a business.

British Petroleum (BP) once coined the phrase "Beyond Petroleum" as part of a corporate re-branding strategy. As BP exited from solar power in 2011 and wind energy in 2013, the firm has returned to its original name. As BP navigates away from the Deep Horizon disaster, there are many investors in the City of London that are wondering what the future holds for organizations that do not learn from mistakes of the past.

9

WEATHER AND ENERGY

INDIA, 2012

INDIA IS THE third largest consumer and producer of electricity in the world after the U.S. and China. With a population of over 1.2 billion people, the country suffers from a fragile power system, and 300 million people have no access to electricity. The summer of 2012 was particularly hot and demand hit record highs in New Delhi. Late arrival of the monsoons meant that hydropower production was lower than usual and agricultural areas had increased demand for operating irrigation pumps to paddy fields. As a result, on July 30, circuit breakers on the 400 kV Bina-Gwalior line tripped and power failures eventually cascaded through the grid, leaving more than 300 million people without electricity. The next day the system failed again and over 620 million people, almost half of the population of India, had no electricity. This blackout was the largest power outage in history with an estimated 32 gigawatts of generating capacity being taken offline and affecting about 9 percent of the world population. The application of data, for insight, could have better prepared the country or perhaps even prevented the blackout.

THE WEATHER

The weather serves as an interesting case study for understanding the paradigm shifts that are currently taking place in different industrial sectors. There is a growing realization that data and appropriate analytics can be used to monitor, describe, and forecast the evolution of key performance indicators. There is no doubt that there are considerable technical challenges in moving from forecasts of physical systems to socioeconomic systems — or indeed eventually being in a position to understand human behavior.

As most big data applications involve the fusion of two or more different sources of information or disciplines, we will focus on the relationships between weather and renewable energy. It is also interesting to understand just how important the communication of the outputs from data analytics are when being used for critical decision-making.

FORECASTING THE WEATHER

Throughout time, there have been many approaches used for forecasting the weather. Farming communities have noticed that animals display certain characteristics before particular weather events. A famous old wives' tale suggests that a field full of cows lying down indicates a rainstorm is coming.

The Babylonians first attempted to use cloud patterns to forecast the weather in 650 BC. Aristotle wrote *Meteorologica* in 340 BC, which contains theories about weather phenomena such as rain, wind, clouds, hail, thunder, lightning, and hurricanes. It was not until the Renaissance period that Italian scientists invented instruments to measure the actual properties of the atmosphere. Important inventions such as the thermometer (by Galileo Galilei around 1592) for measuring temperature and the barometer (by Evangelista Torricelli in 1643) for measuring atmospheric pressure finally offered a means of observing and recording the weather.

Two Royal Navy officers, Francis Beaufort and Robert FitzRoy, are widely credited with transforming weather forecasting into a science. Their work led to the development of reliable tide tables and the collection of weather data at sea by ships' captains as a service to mariners. It was the loss of the Royal Charter in 1859 due to a storm that inspired FitzRoy to develop charts for forecasting the weather.

By the 20th century, scientists had gained a deeper understanding of the atmosphere and were in a position to propose a mathematical model for describing how it changes over time. In 1922, the English scientist Lewis Fry Richardson published "Weather Prediction by Numerical Process," based on notes he had written while working as an ambulance driver during World War I.

As Richardson's approach to numerical weather prediction came before computers were invented, he envisioned a large auditorium of thousands of people performing the calculations and passing them to others (see the following figure). It is not such a leap of the imagination to view this as the forerunner of modern data analytics.

Richardson's forecast factory

http://celebrating200years.noaa.gov/foundations/numerical_wx_pred/theater.html

WHEN ARE WEATHER FORECASTS WRONG?

Weather forecasts certainly have their limitations. Yet is it easier to produce forecasts than it is to undertake a serious scientific evaluation of their quality. Nobody assumes that weather forecasts will always be correct, but exactly how much error should we tolerate and how often should we expect to be let down? Forecast performance should be measured relative to the natural variability of the weather data of interest — for example, forecasting the amount of rain falling in the Sahara desert is a relatively easy challenge.

The models utilized for numerical weather prediction suffer from insufficient observations across the globe and a lack of computational resources for propagating this initial information to future forecast horizons. Our view of how well a weather forecast service is doing is typically based on our memory of events associated with failure. If we use weather forecasts to make decisions and our actions lead to financial gains or penalties, then we can perform a cost-benefit analysis of a given forecasting service. Public attention is at its peak around the time of extreme events and therefore a forecaster's ability to correctly warn of such extremes is perceived as being important.

Michael Fish was a household name in the U.K., a popular weather presenter for the British Broadcasting Corporation (BBC). Unfortunately, his words of reassurance on October 15, 1987, are associated with one of the greatest forecast errors in the U.K. He started by saying "Earlier on today apparently a woman rang the BBC to say she'd heard there was a hurricane on the way," and then continued with, "Well, if you're watching, don't worry — there isn't." Although not technically a hurricane, what followed was the worst storm to hit southeast England since 1703. This particular forecast error made the weather forecaster infamous as the man who failed to spot what became known as the Great Storm.

CHAOS

It is, of course, easy to throw tomatoes at the scientists responsible and blame them for their inadequate modeling. But what should we realistically expect from the models? Chaos theory plays an important role here, as the atmosphere behaves like a chaotic system — small initial errors tend to be amplified as the forecast horizon is increased. Ed Lorenz, the meteorologist who discovered chaos in weather models, famously speculated that a butterfly flapping its wings in Brazil could eventually cause a tornado in Texas. So should we give up on forecasting the weather because of chaos? Definitely not. Uncertainty is something that all data scientists have to live with, but the methods used for accounting for uncertainty vary dramatically, as explained in the next section.

ENSEMBLE FORECASTS

Dealing with uncertainty in the *initial condition,* the starting value for numerical weather prediction (NWP), is relatively straightforward. Indeed, multiple weather service providers, such as the European Centre for Medium-Range Weather Forecasts (ECMWF) and the U.S.'s National Centers for Environmental Prediction (NCEP), use ensemble weather forecasts to manage this uncertainty. An *ensemble weather forecast* is obtained by running the NWP model multiple times to obtain a collection of different scenarios, each originating with a different initial condition. Now, the whole point of generating ensemble forecasts is to provide an improved forecasting approach and a better way of communicating to end-users.

Unfortunately, clear communication has not always been the result of using ensemble forecasts. Indeed, the U.K.'s national weather service, known as the Met Office, has been haunted by its use of seasonal weather forecasts. By definition, *seasonal forecasting* refers to horizons of up to six months ahead, and this is pushing the boundaries of what can be accurately predicted given current technical capability and computational resources. Seasonal forecasts use an ensemble approach to evaluate whether future temperatures are likely to be lower than, equal to, or higher than typical values.

The Met Office ran into difficulties in 2009 when it forecasted a "barbecue summer" that did not materialize. In January 2011, its "mild winter" ended up being the coldest in 31 years. As a result of the controversy around these seasonal forecasts, the BBC launched an attack on the U.K. Met Office, with a BBC *Newsnight* investigation into their failure to forecast the extremely cold conditions. The Met Office's model stated that there was a 50-percent probability of a "mild winter," and this was the most likely outcome of the three possible outcomes. The media sound bite from the Met Office presented this most likely outcome as the forecast and failed to discuss the other possibilities. In retrospect, it seemed unfair that the Met Office got such bad publicity, and it was arguably the communication that was at fault, rather than the model. Later that year, however, the Met Office scrapped its seasonal forecasts after carrying out extensive customer research.

COMMUNICATION

The previously mentioned challenge faced by the Met Office offers an important lesson for all those hoping to reap the benefits of big data. Analytics, by their very nature, always depend on a mathematical model, and this model will often be wrong due to observational uncertainty, difficulties in estimating parameters, and failure to identify the correct model structure. Therefore, the mechanism delivering the analytics is uncertain, and the method used to communicate this uncertainty is crucial for building confidence. If forecasters are honest, they should express their forecasts using an expected value and a "confidence interval," to indicate the range of possible outcomes. Forecasters who have no confidence intervals are claiming perfect precision, which certainly does not imply that their forecasts are more accurate, but more likely that they are ignorant about potential forecast errors. Unfortunately, human nature is such that the overconfident forecasters are sometimes viewed as having a better approach or simply knowing more than their competitors. Deploying statistical techniques to evaluate historical forecasts is the only way to figure out which forecaster is best.

RENEWABLE ENERGY
SOLAR, HYDRO, AND WIND POWER

In an attempt to address the challenge posed by climate change, many countries are setting stringent targets for renewable energy generation over the next decade to hopefully reduce fossil fuel consumption. The European Union (E.U.) is committed to generating 20 percent of its energy from renewable sources by 2020. In addition, a strong desire for energy security is driving many countries towards renewable energy. At present, over one-fifth of the world's electricity is produced from renewable sources. The penetration of various renewable energy sources such as wind, solar, hydro, wave and tidal power, geothermal energy (heat from the Earth) and biomass (fuel from organic materials) is increasing in many countries. While this approach to generating electricity is at the heart of the green economy, there are many technical and operational challenges that must be overcome.

In the good old days, when fossil fuel sources were used to generate electricity, it was relatively easy for power-system operators to plan ahead and balance supply and demand. It is true that demand was always varying throughout the day, but at least most generators could be switched on and off, and their capacities were known and fixed. In contrast, wind energy was one of the first sources of renewable energy to introduce a variable form of supply into the power system. When I spoke with a system operator a few years ago, there was relief in his voice as he said he was retiring and would not have to worry about wind-power variability anymore.

It is true that renewable energy adds substantial complexity to the power system, but this does not mean that the system cannot be adapted to cope with variable sources of electricity. As with many systems for which the competition for resources is increasing, it makes sense to ensure that appropriate data is being collected so that intelligent analytics can be brought to bear on the problems. Many wind farms consist of multiple turbines, but it is not always known how many turbines are in operation at any given time due to maintenance issues or equipment failure. This lack of local operational data makes it difficult to understand the relationship between wind speed and wind power being generated at the level of the wind farm. The lesson to be learned for anyone wishing to utilize big data in their business, is the importance of ensuring that the data collection mechanism is appropriate to enable construction of a decision-support tool.

Renewable energy generation is now of such importance in some countries that this particular component of the available supply resource can often be responsible for determining the electricity prices in the market. Therefore, accurate weather forecasts for wind, rain, luminosity, and temperature are required for determining electricity prices. Indeed, some hedge funds have used this relationship between weather forecasts and price to construct profitable trading strategies.

VOLATILE OR INTERMITTENT SUPPLY

The specific term used to describe the supply of renewable energy is important, as it conveys a view about its potential role. The word *intermittent* has negative connotations, as it suggests that the electricity supply simply switches on and off. While this might be true for an individual wind farm — some days have no wind and therefore no power being generated — the supply also varies over time when the turbines are rotating. It is more accurate to describe wind power generation as being *volatile*. Indeed, there is a diversification effect that implies the aggregate electricity generated by a collection of wind farms scattered across a country is unlikely to be zero because of the different wind power profiles that are produced at these sites. Furthermore, by combining a range of different renewable energy supplies, such as wind, tidal and solar that depend on multiple environmental factors, the aggregate electricity is less variable over time. For this reason, it is important to carefully plan the locations of future sources of generation and to consider the impact on the aggregate supply.

Big data enables us to observe complex systems in greater detail, understand what factors are relevant, and improve basic scientific theories. The relationship between wind speed and wind power follows a nonlinear curve but has a cut-in point when the wind speeds is too low to generate energy and a cut-off point when the wind speeds are so high as to damage the turbine. In theory, according to the laws of physics, the relationship for an individual wind farm should follow a cubic power law, suggesting that a relatively small rise in wind speed produces a large increase in wind power. However, big data offers a means of analyzing empirical observations to show that there are substantial deviations from theory because of the time of year and local terrain at the site of the wind farm.

Similarly, the theory that describes how much electricity should be produced by photovoltaic technology rarely matches reality. The ideal environmental conditions used to test the technologies are rarely met in practice. Luckily, this is where empirical data can be useful in deciding which technology is optimal for particular environmental conditions and geographical location. At Oxford University, we measured the efficiency of 11 different photovoltaic technologies over a couple of years in both sunny Mallorca in Spain and cloudy Oxford in the U.K. and our data analysis allowed us to estimate what could realistically be expected in a given geographical location.

Atmospheric models and ensemble weather forecasts are invaluable in the renewable energy sector for a variety of reasons. Atmospheric models can be employed to generate consistent data sets for different weather variables, such as temperature and wind speed, that span the globe. This data can then be used to provide risk assessments for physical infrastructure and to evaluate performance. Ensemble forecasts provide not only accurate estimates of future production but also quantify the uncertainty in the expected production. In addition, the ensembles provide the basis of a probabilistic forecasting system and can be used to provide early warnings of extreme events such as sudden ramps in the amount of power being generated. According to James Taylor, Patrick McSharry & Roberto Buizza ("Wind power density forecasting using ensemble predictions and time series models," *IEEE Transactions on Energy Conversion*, 2009), these probabilistic forecasts can be accurate for up to eight days ahead and can be used to facilitate participation in the electricity market.

ENERGY CONSUMPTION

Pretty much everything we do today relies on our ability to consume energy. At home and at work, we use equipment and services that require electricity. Electricity demand therefore provides a profile of human behavior. Most countries collect electricity-demand data every half hour or every 15 minutes, and the shape of the curve throughout the day shows when we get up in the morning, when we work, and when we come home in the evening. Along-side seasonality due to the time of day and day of the week, there are other obvious patterns such as bank holidays and vacations. Power system operators also spend considerable time worrying about important sporting events such as when national teams play in the FIFA World Cup. At half time, England supporters tend to switch on their electric kettles (over 500,000 kettles) to have a cup of tea, and this synchronized spike of demand causes a real headache for those in charge of balancing supply and demand at the U.K. National Grid. After all, an electricity black-out during a World Cup match would not be appreciated.

The regularity of our daily activities and the collective actions of a nation tend to generate peaks in demand at particular times of the day. If only we were more diverse in our behaviors, it might be easier to smooth out electricity demand and avoid these peaks.

Of course, we humans are only part of the story. The weather enters again here to cause more problems. When it gets very cold, we tend to switch on heating devices, and therefore the demand increases. During the summer, in order to cope with high temperatures, air conditioning units are turned on. These two activities cause a V-shaped pattern in demand versus temperature. Other environmental variables should also be considered, such as humidity and wind chill, as it is our perception of the effective temperature that leads us to adjust our thermostats.

Faced with limited sources of energy and a desire to decrease carbon emissions, it makes sense to ask how could we use big data to maintain our quality of life and yet modify our consumption of energy to accomodate more renewable sources. If we can measure and forecast both electricity supply and demand, surely there is a way to construct intelligent mechanisms to facilitate this change.

SMART METERS

The first step is to actually find out what is going on in our homes. We might have some idea of our electricity consumption on a monthly or quarterly basis, but few people know what happens each day. Even fewer people know which devices are responsible for the greatest usage. Smart meters are slowly being introduced into homes around the world in the hope of causing a paradigm shift in the way we consume resources such as electricity, gas, and water. Plans for introducing smart meters in different countries vary greatly. For example, the U.K.'s Department of Energy and Climate Change wishes to have smart meters installed in all homes and businesses by 2020, motivated by its impact assessment which forecasts a potential net benefit of £6.2 billion.

Smart meters will certainly generate some interesting data, as each home could produce a data point every half hour — or even more frequently — instead of monthly, and this information will be fed back to the utility for monitoring and billing. According to market analyst GTM Research, global utility company expenditure on big data analytics is expected to grow from $700 million in 2012 to $3.8 billion in 2020. For the utilities, smart meters will offer many benefits such as, an ability to improve forecasts of demand at a household level, predictions of outages, and a way to identify leaks and potential fraud.

Various instruments are now available to measure your electricity consumption and, after much experimentation through switching every device in the house off and on, it is often surprising to find out which devices suck up most of the electricity. Of course, it is important to consider both the power used and the typical duration to make decisions about reducing consumption. It will be much easier to obtain this information once a usable dashboard is available to monitor the consumption of each electrical device in the house. For example, Navetas Energy Management offers technology solutions to help individuals and companies make better energy choices.

Being able to monitor and understand is surely the precursor to seeing what changes can be made and then finally taking action. Again, there is much scope for developing intelligent machines to automatically control our electricity demand within the home. Utilities could offer different prices throughout the day for energy usage, which would encourage us to allocate our consumption effectively. It would be extremely tedious to have to make these consumption changes manually. On the other hand, once smart meters become widespread, it would be possible to develop an intelligent machine that deploys big data to understand the behavior and patterns of the residents, the variation in electricity prices, and the relative importance of powering and charging different devices throughout the day.

The introduction of smart meters is just one part of the overall revolution behind the *smart grid*, which refers to an electrical grid driven by the use of digital technology to collect data about both supply and demand and make decisions to improve key performance indicators, such as the efficiency, reliability, and sustainability of the power system. The operation of this smart grid relies on data collected from meters and sensors about cables, transformers, overhead lines, substations, and the activities of engineers and customers.

INTELLIGENT DEMAND-SIDE MANAGEMENT

Demand-side management refers to the process of encouraging customers to change the way they consume electricity. Due to the technical challenge of meeting high demand, the objective of demand-side management is to encourage customers to use less electricity during peak hours or to shift their demand to off-peak periods such as nighttime and weekends. Typical approaches rely on financial incentives, rewards, and education.

Providing an effective nudge to encourage behavior change is an active area of research known as *behavioral economics*. Traditional economics has tended to focus on the use of money to motivate behavior change via financial incentives and fines. Academic researchers have found that individuals are greatly influenced by knowing what their neighbors and peers are consuming. Usually, the goal of demand-side management is to encourage the consumer to use less energy during peak hours or to move the time of energy use to off-peak times, such as nighttime and weekends.

Once sufficient numbers of smart meters are installed, it will be possible to collect big data relating to intraday consumption patterns at the household level, local weather conditions, variation in electricity prices, and socioeconomic information about individuals and their preferences. Putting all of this data together will provide a means of constructing intelligent demand-side management approaches in which a machine makes the necessary decisions to satisfy the preferences of the household members while also helping to improve the sustainability of the power system.

INDIA, 2012 (CONTINUED)

In response to the blackout in India, the United States Agency for International Development (USAID) proposed that another widespread outage could be prevented by creating an integrated network of microgrids and distributed generation connected seamlessly with the main grid using smart grid technology. The current fragility of power systems is a challenging issue in both developed and developing countries. Extreme rain and wind caused the 2009 Brazil and Paraguay blackout that affected at least 60 million people. In California, the Great Blackout of 2011, caused by a technician mistakenly cutting a line between two substations, left seven million people without electricity.

The weather will have an increasingly important role in both governing the supply of renewable energy and driving our demands for electricity as we seek to maintain comfortable living conditions despite fluctuations in environmental variables. Our addiction to consuming increasing quantities of energy is unlikely to decline as a large portion of the world's population continues to develop. In order to guarantee security of supply, we will need to adapt to changing weather patterns and find ways of remaining resilient to potential weather extremes that may become more commonplace as a result of a changing climate. Our ability to adapt will likely involve a compromise between easy fixes based on technological solutions and less attractive behavior change. Big data and quantitative modeling will play a large part in the technological innovations that will be necessary to promote sustainability and make sure that the lights stay on.

II

LEARNING FROM PATTERNS IN BIG DATA

10 PATTERN RECOGNITION

ELEMENTS OF SUCCESS RHYME

THE ROLE THAT data will play in defining and transforming industries over the next decade is evident. Whether it's reinventing the practice of farming or redefining the role of a retailer, data drives the revolution. The stories in Part 1 of this book were chosen because they demonstrate early signs of the data revolution. However, there are countless other industries and businesses that will also lead, be impacted by, or suffer from the data revolution. The purpose of Part 2 is to help prepare anyone, in any business, for the coming revolution.

The science of pattern recognition has been explored for hundreds of years, with the primary goal of optimally extracting patterns from data or situations, and effectively separating one pattern from another. Applications of pattern recognition are found everywhere, whether it's categorizing disease, predicting outbreaks of disease, identifying individuals (through face or speech recognition), or classifying data. In fact, pattern recognition is so ingrained in many things we do, we often forget that it's a unique discipline which must be treated as such if we want to really benefit from it.

According to Tren Griffin, a prominent blogger and IT executive, Bruce Dunlevie, a general partner at the venture capital firm Benchmark Capital, once said to him, "Pattern recognition is an essential skill in venture capital." Griffin elaborates the point Dunlevie was making that "while the elements of success in the venture business do not repeat themselves precisely, they often *rhyme*. In evaluating companies, the successful VC will often see something that reminds them of patterns they have seen before." Practical application of pattern recognition for business value is difficult. The great investors have a keen understanding of how to identify and apply patterns.

Jeff Bezos, founder and CEO of Amazon.com, has famously stated that companies should base their strategy on things that won't change. His point is that many companies build a strategy based on unreliable variables, and when those variables change, the strategy is out of date. For example, if a retailer bases their strategy on identifying and bringing to market the hippest designs, their strategy will inevitably fail. While they will have years of success (when they choose the right designs), they will have other years where they placed the wrong bet. Therefore it's not a sustainable strategy.

In contrast, Bezos has based Amazon's strategy on offering wide selection, low prices, and fast, reliable delivery. His point is that no matter what changes in retail or in the world, consumers will always want low prices. They will always appreciate wide selection. And once they choose an item, they will want it quickly. He has based the strategy on the things that he knows will *not* change. This is a modern application of pattern recognition.

PATTERN RECOGNITION: A GIFT OR TRAP?

Written in 2003 by William Gibson, *Pattern Recognition* (G.P. Putnam's Sons) is a novel that explores the human desire to synthesize patterns in what is otherwise meaningless data and information. The book chronicles a global traveler, a marketing consultant, who has to unravel an Internet-based mystery. In the course of the book, Gibson implies that humans find patterns in many places, but that does not mean that they are always relevant. In one part of the book, a friend of the marketing consultant states, "Homo sapiens are about pattern recognition. Both a gift and a trap." The implication is that humans find some level of comfort in discovering patterns in data or in most any medium, as it helps to explain what would otherwise seem to be a random occurrence. The trap comes into play when there is really not a pattern to be discovered because, in that case, humans will be inclined to discover one anyway, just for the psychological comfort that it affords.

Patterns are useful and meaningful only when they are valid. The bias that humans have to find patterns, even if patterns don't exist, is an important phenomenon to recognize, as that knowledge can help to tame these natural biases.

WHAT FISH TEACH US ABOUT PATTERN RECOGNITION

Pattern recognition is both an art and a science. Historically, the roots of pattern recognition are based in science, with pioneers like Thomas Bayes leading research on the topic.

BAYES' THEOREM

Thomas Bayes died in 1761. At his death, he left behind two unpublished essays that aimed to determine probabilities of causes from observed effects. As was typical for the times, the essays were sent to the British Royal Society, where they were undoubtedly filed and quickly forgotten.

Many years later, a French mathematician, Pierre-Simon Laplace, pursued a similar discovery around probabilities and effects. Hearing of his work, the English quickly claimed this discovery as their own, citing what would become known as Bayes' Theorem.

Bayes' Theorem provides a statistical approach to the problem of pattern classification. The approach demands that in any situation, the tradeoffs between a variety of decisions are quantified, using the probability and the costs associated with each decision. Exploiting prior knowledge and observing what's different (or not) is critical to the process.

Bayesian Probability Theory is purely a statistical approach, which assumes that the underlying probabilities are known perfectly. This leads to theoretically optimal decisions. But, unfortunately, the world is not always so simple.

TSUKIJI MARKET

The seafood will start arriving at Tsukiji before four in the morning, so an interested observer must start his day quite early. The market will see 400 different species passing through on any given day, eventually making their way to street carts or the most prominent restaurants in Tokyo. The auction determines the destination of each delicacy.

In any given year, the fish markets in Tokyo will handle over 700 metric tons of seafood, representing a value of nearly $6 billion.

The volume of species passing through Tsukiji represents an interesting challenge in organizing and classifying the catch of the day. In the 2001 book *Pattern Classification* (Wiley), Richard Duda provided an interesting view of this process, using fish as an example.

Pattern Classification

With a fairly rudimentary example — fish sorting — Duda is able to explain a number of key aspects of pattern recognition.

A worker in a fish market, Tsukiji or otherwise, faces the problem of sorting fish on a conveyor belt according to their species. This must happen over and over again, and must be done accurately to ensure quality. In Duda's simple example in the book, it's assumed that there are only two types of fish: sea bass and salmon.

As the fish come in on the conveyor belt, the worker must quickly determine and classify the fishes' species.

There are many factors that can distinguish one type of fish from another. It could be the length, width, weight, number and shape of fins, size of head or eyes, and perhaps the overall body shape.

There are also a number of factors that could interrupt or negatively affect the process of distinguishing (sensing) one type from the other. These factors may include the lighting, the position of the fish on the conveyor belt, the steadiness of the photographer taking the picture, and so on.

The process, to ensure the most accurate determination, consists of capturing the image, isolating the fish, taking measurements, and making a decision. However, the process can be enhanced or complicated, based on the number of variables. If an expert fisherman indicates that a sea bass is longer than salmon, that's an important data point, and length becomes a key feature to consider. However, a few data points will quickly demonstrate that while sea bass are longer than salmon on average, there are many examples where that does not hold true. Therefore, we cannot make an accurate determination of fish type based on that factor alone.

With the knowledge that length cannot be the sole feature considered, selecting additional features becomes critical. Multiple features — for example, width and lightness — start to give a higher-confidence view of the fish type.

Duda defines *pattern recognition* as the act of collecting raw data and taking an action based on the category of the pattern. Recognition is not an exact match. Instead, it's an understanding of what is common, which can be expanded to conclude the factors that are repeatable.

PATTERN RECOGNITION

Pattern recognition, with elements of science and art, can be taught. The process, the approach, and the outcomes can all be learned. However, the most difficult part is learning how to apply it in a given situation.

ROCHESTER INSTITUTE OF TECHNOLOGY

The Rochester Institute of Technology (RIT) sits on the banks of the Genesee River. RIT is often thought of for its prowess in computer science and technology. Richard Zanibbi, a professor in the Department of Computer Science, defines *pattern recognition* as the identification of implicit objects, types, or relationships in raw data by an animal or machine. Said more simply, it's recognizing hidden information in data. He goes further to delineate the process associated with pattern recognition, which is designed to answer three questions:

- **What is it?** This is the task of *classification*.
- **Where is it?** This is the task of *segmentation*.
- **How is it constructed?** This is the collection of tasks to parse data and recognize the patterns. Said another way, this is determining the shape of the data; understanding how the data is related and how it forms an expression.

A METHOD FOR RECOGNIZING PATTERNS

Answering the three key questions (what is it?, where is it?, and how it is constructed?) seems straightforward — until there is a large, complex set of data to be put through that test. At that point, answering those questions is much more daunting. Like any difficult problem, this calls for a process or method to break it into smaller steps. In this case, the method can be as straightforward as five steps, leading to conclusions from raw inputs:

1. **Data acquisition and sensing:** The measurement and collection of physical variables.
2. **Pre-processing:** Extracting noise in data and starting to isolate patterns of interest. In the fish example given earlier in the chapter, you would isolate the fish from each other and from the background. Patterns are well separated and not overlapping.

3. **Feature extraction:** Finding a new representation in terms of features. For the fish, you would measure certain features.

4. **Classification:** Utilizing features and learned models to assign a pattern to a category. For the fish, you would clearly identify the key distinguishing features (length, weight, etc.).

5. **Post-processing:** Asessing the confidence of decisions, by leveraging other sources of information or context. Ultimately, this step allows the application of content-dependent information, which improves outcomes.

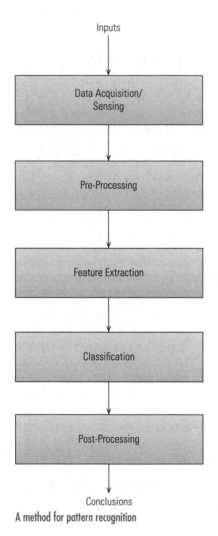

A method for pattern recognition

Pattern recognition techniques find application in many areas, from machine learning to statistics, from mathematics to computer science. The real challenge is practical application. And to apply these techniques, a framework is needed.

Clay Christensen

Clay Christensen is the Kim B. Clark Professor of Business Administration at the Harvard Business School. The defining moment of his career thus far is the publication of *The Innovator's Dilemma: When New Technologies Cause Great Firms to Fail* (Harvard Business Review Press, 1997), an assessment of why many technology companies fail to maintain a leadership position after initial success in a market. He coined the term *disruptive innovation* to refer to the phenomena whereby a new entrant to the market disrupts the incumbent by leaping a few steps ahead in terms of cost and/or technology advantage. This happens largely because the incumbent, based on their initial success, becomes a prisoner to their customers and can offer only incremental (not disruptive) innovation after their initial innovation.

Christensen often cites the steel industry as an example of disruptive innovation, as he believes the patterns present there offer a good lesson on disruptive innovation. The story begins when there were only a few dominant steel mills in the country, as they were the only companies with the scale and investment to produce high-quality steel. In time, mini-mills began to emerge that made steel only from scrap. So, by definition, the quality of the steel produced was very poor. This low-quality steel was known as rebar (reinforcing bar), and it was commonly understood that nearly anyone could cheaply produce this low-quality steel. The integrated mills had no interest in this market segment, as they saw their competitive advantage as their ability to produce high-quality steel, the kind of steel used for automobiles or appliances. Plus, the high-quality steel was higher in profit margin, and therefore the integrated mills were happy to see the mini-mills produce the low-profit and low-quality steel.

Eventually, the price of rebar collapsed. As the integrated mills exited the production of rebar, the mini-mills began an aggressive price competition, and nearly overnight, prices were driven down by 20 percent. Suddenly, rebar was no longer attractive to the mini-mills. So they were forced to look upmarket, and they started to attack the next tier of steel, which was being produced by the integrated mills. From this point, the story repeats: The integrated mills no longer saw that tier as attractive or profitable and were happy to see the mini-mills assume that part of the market. Eventually, prices collapsed in that part of the market, too, and as the mini-mills moved upmarket again, the integrated mills quickly found themselves out of business. This is the vicious cycle of disruptive innovation.

With *The Innovators Dilemma* and his subsequent works, Christensen does not offer an answer to the problems discussed. In fact, he is very clear on this point, as the answer will be different for every company and industry, depending on the factors at play. Instead, Christensen offers a framework for decision-making. He has always asserted that his goal is to teach others how to think about the problem and solution, instead of trying to provide the solution.

The notion of pattern recognition must be considered similarly to how Christensen offers council on business strategy challenges. There is no single pattern or answer for an industry, company, or individual. The patterns will differ based on the dynamics of each situation. However, the purpose of this chapter is to share a perspective on how to think about pattern recognition in order to apply it in any situation. The methodology offers five steps to consider, but there could be more or less in any given situation.

In the case of farming and agriculture, pattern recognition is being used more frequently. Plant disease is one of the most important detractors of farm productivity, causing the destruction and loss of crops. Detecting disease at an early stage provides the opportunity to effectively treat the condition and avoid the loss of productivity. Early detection puts the impetus on developing a system (or application) for classifying disease in infected plants, including analysis of the color, shape, texture, or other factors. There is no single answer to stemming disease in plants, as the variables are different in each situation. However, the right methodology, consistently applied, can become an early warning system.

ELEMENTS OF SUCCESS RHYME (CONTINUED)

Pattern recognition can be a gift or a trap.

It's a trap if a person is lulled into believing that history repeats itself and therefore there is simply a recipe to be followed. This is lazy thinking, which rarely leads to exceptional outcomes or insights.

On the other hand, it's a gift to realize that, as mentioned in this chapter's introduction, the elements of success rhyme. Said another way, there are commonalities between successful strategies in businesses or other settings. And the proper application of a framework or methodology to identify patterns and to understand what is a pattern and what is not can be very powerful.

The inherent bias within humans will seek patterns, even where patterns do not exist. Understanding a pattern versus the presence of a bias is a differentiator in the Data era. Indeed, big data provides a means of identifying statistically significant patterns in order to avoid these biases.

11

WHY PATTERNS IN BIG DATA HAVE EMERGED

MEATPACKING DISTRICT

A WALK THROUGH the meatpacking district in New York City is a lively affair. In the early 1900s, this part of town was known for precisely what its name implies: slaughterhouses and packing plants. At its peak, there were nearly 300 such establishments, located somewhat central to the city and not far from shipping ports. Through the years, this area of town has declined at times, but in the early 1990s, a resurgence began that shows no signs of ending.

Located in some proximity to the fashion district of Manhattan, the Meatpacking district stands for what is modern, hip, and trendy; a bastion of culture in the work-like city. Numerous fashionable retailers have popped up, along with trendy restaurants. And, on the fringes, there is evidence of the New York startup culture flowing from the Flatiron District, sometimes known as Silicon Alley. A visit to one of the companies in this area opened my eyes to some of the innovation that is occurring where data is the business, instead of being an enabler or addition to the business.

As I entered the office of this relatively newborn company, I was confident that I understood their business. It was pretty simple: They were a social sharing application on the web that enabled users to easily share links, share content, and direct their networks of friends to any items of interest. The term *social bookmarking* was one description I had heard. The business seemed straightforward: Attract some power users, enable them to attract their friends and associates, and based on the information shared, the company could be an effective ad-placement platform, since it would know the interests of the networks and users.

But what if social bookmarking and ad placement was not the business model at all? What if all that functionality was simply a means to another end?

BUSINESS MODELS IN THE DATA ERA

Data has the power to create new businesses and even new industries. The challenge is that there are many biases about the use of data in a business. There is a view that data is just about analytics or reporting. In this scenario, it's relegated to providing insight about the business. There is another view that data is simply an input into existing products. In this case, data would be used to enrich a current business process, but not necessarily change the process. While these cases are both valid, the power of the Data era enables much greater innovation than simply these incremental approaches.

There are three classes of business models for leveraging data:

- **Data as a competitive advantage:** While this is somewhat incremental in its approach, it is evident that data can be utilized and applied to create a competitive advantage in a business. For example, an investment bank, with all the traditional processes of a bank, can gain significant advantage by applying advanced analytics and data science to problems like risk management. While it may not change the core functions or processes of the bank, it enables the bank to perform them better, thereby creating a market advantage.

- **Data as improvement to existing products or services:** This class of business model plugs data into existing offerings, effectively differentiating them in the market. It may not provide competitive advantage (but it could), although it certainly differentiates the capabilities of the company. A simple example could be a real estate firm that utilizes local data to better target potential customers and match them to vacancies. This is a step beyond the data that would come from the Multiple Listing Service (MLS). Hence, it improves the services that the real estate firm can provide.

■ **Data as the product:** This class of business is a step beyond utilizing data for competitive advantage or plugging data into existing products. In this case, data is the asset or product to be monetized. An example of this would be Dun & Bradstreet, which has been known as the definitive source of business-related data for years.

In these business models, there are best practices that can be applied. These best practices are the patterns that teach us how to innovate in the Data era.

DATA AS A COMPETITIVE ADVANTAGE

Procter & Gamble is legendary for the discipline of brand management. To be the leading consumer packaged-goods company in the world, brand is everything. Ultimately, consumers will buy brands that they know and understand, and those that fulfill an expectation. This was one of the first lessons that Scott Cook learned when he joined Procter & Gamble in the 1970s. He also observed that the core business processes for managing a brand (accounting, inventory, etc.) would be transformed and subsumed by the emerging personal computer revolution. This insight led him to co-found Intuit in 1983, really at the dawn of the expansion of personal computers. The premise was simple: Everyone, small companies and individuals alike, should have access to the financial tools that previously had been reserved for large enterprise companies.

Now broadly known for tax preparation software (TurboTax), along with software solutions for small businesses (QuickBooks) and individuals (Quicken), Intuit has transformed many lives. At over $4 billion in revenue and nearly 8,000 employees, the company is a success by any barometer. Intuit is one of the few software companies to challenge Microsoft head-on and not only live to tell about it, but to prosper in its wake. Microsoft Money versus Quicken was a battle for many years; Microsoft was trying to win with software, while Intuit was focused on winning with data, utilizing software as the delivery vehicle.

Robbie Cape, who ran Microsoft's Money business from 1999 to 2001, believes that Intuit's advantage had very little to do with technology. Instead, he attributes Intuit's success to its marketing prowess. While there may be some truth to the statement, its very hard to believe that Intuit had deep-enough pockets to out-market Microsoft. Instead, the differentiation seems to come from data.

The NPD Group analyst Stephen Baker said that Intuit won by building out critical mass in financial software and the surrounding ecosystem. Intuit had the insight that adjacency products and services, leveraging data, made the core software much more attractive. This insight led to their early and sustained domination of the retail channel.

Intuit's ability to collect and analyze a large amount of sensitive and confidential data is nearly unsurpassed. Nearly 20 million taxpayers use TurboTax online, sharing their most personal data. Over 10 million customers use QuickBooks, with employee information for over 12 million people flowing through its software. Brad Smith has been cited as declaring that 20 percent of the United States Gross Domestic Product flows through QuickBooks. No other collection of data has this type and extent of financial information on individuals and small businesses.

With these data assets, Intuit began to publish the Intuit Small Business Index. The Index provides summarized insights about sales, profit, and employment data from the small businesses that use QuickBooks. This information can provide headlights to business and salary trends, which ultimately becomes insight that can be fed back into the product. This was the point that Microsoft Money either missed or simply could not achieve: The value was never in the software itself. The value was in the collection, analysis, and repurposing of data to improve the outcomes for the users.

In 2009, Intuit purchased Mint, a free web-based personal-finance application. Mint took the Intuit business model a step further: They provide their software for free, knowing that it's a means to an end. The social aspects of Mint enable users to do much more than simply track their spending. Instead, it became a vehicle to compare the spending habits of an individual to others of a similar geography or demographic. The user can do these comparisons, or the comparisons can show up as recommendations from Mint. Further, Mint brings an entirely different demographic of data to Intuit. While the Intuit customer base was largely the 40-and-over demographic (people who had grown up with Quicken, QuickBooks, etc.), Mint attracted the Millennial crowd. The opportunity to combine those two entirely different sets of data was too attractive for Intuit to pass up.

To date, Intuit has not had a strategy for monetizing the data itself. Perhaps that may change in the future. However, with data at the core of its strategy, Intuit has used that data to drive competitive advantage, while software was merely the delivery vehicle. The companies that tried the opposite have not fared so well.

DATA IMPROVES EXISTING PRODUCTS OR SERVICES

Chapters 1 through 9 offer a multitude of examples in which data is being utilized to improve existing products or services. In the pursuit of a business model leveraging data, this category is often the low-hanging fruit; more obvious, although not necessarily easy to do. The examples covered previously are:

- **Farming and agriculture:** Monsanto is using data to augment applications like FieldScripts, which provides seeding prescriptions to farmers based on their local environments. While Monsanto could provide prescriptions through their normal course of operation, data has served to personalize and thereby improve that offering.
- **Insurance:** Dynamic risk management in insurance, such as pay-as-you-drive insurance, leverages data to redefine the core offering of insurance. It changes how insurance is assessed, underwritten, and applied.
- **Retail and fashion:** Stitch Fix is redefining the supply chain in retail and fashion, through the application of data. Data is augmenting the buying process to redefine traditional retail metrics of inventory, days sales outstanding, etc.
- **Customer service:** Zendesk integrates all sources of customer engagement in a single place, leveraging that data to improve how an organization services customers and fosters loyalty over time.
- **Intelligent machines:** Vestas has taken wind turbines — previously regarded as dumb windmills — and turned them into intelligent machines through the application of data. The use of data changes how their customers utilize the turbines and ultimately optimizes the return on their investment.

Most companies that have an impetus to lead in the Data era will start here: leveraging data to augment their current products or services. It's a natural place to start, and it is relatively easy to explore patterns in this area and apply them to a business. However, it is unlikely that this approach alone is sufficient to compete in the Data era. It's a great place to start, but not necessarily an endpoint in and of itself.

DATA AS THE PRODUCT

Previously in this chapter, the examples demonstrated how data is used to augment existing businesses. However, in some cases, data becomes the product; the sole means for the company to deliver value to shareholders. There are a number of examples historically, but this business model is on the cusp of becoming more mainstream.

DUN & BRADSTREET

In 1841, Lewis Tappan first saw the value to be derived from a network of information. At the time, he cultivated a group of individuals, known as the Mercantile Agency, to act "as a source of reliable, consistent, and objective" credit information. This vision, coupled with the progress that the idea made under Tappan and later under Benjamin Douglass, led to the creation of a new profession: the credit reporter. In 1859, Douglass passed the Agency to his brother-in-law, Robert Graham Dun, who continued expansion under the new name of R.G. Dun & Company.

With a growing realization of the value of the information networks being created, the John M. Bradstreet company was founded in 1849, creating an intense rivalry for information and insight. Later, under the strain caused by the Great Depression, the two firms (known at this time as R.G. Dun & Company and Bradstreet & Company) merged, becoming what is now known as Dun & Bradstreet.

Dun & Bradstreet (D&B) continued its expansion and saw more rapid growth in the 1960s, as the company learned how to apply technology to evolve its offerings. With the application of technology, the company introduced the Data Universal Numbering System (known as D&B D-U-N-S), which provided a numerical identification for businesses of the time. This was a key enabler of data-processing capabilities for what had previously been difficult-to-manage data.

By 2011, the company had gained insight on over 200 million businesses. Sara Mathew, the Chairman and CEO of D&B, commented, "Providing insight on more than 200 million businesses matters because in today's world of exploding information, businesses need information they can trust."

Perhaps the most remarkable thing about D&B is the number of companies that have been born out of that original entity. As the company has restructured over the years, it has spun off entities such as Nielsen Corporation, Cognizant, Moody's, IMS Health, and many others. These have all become substantial businesses in their own right. They are each unique in the markets served and all generate value directly from offering data as a product:

- **Nielsen Corporation:** Formerly known as AC Nielsen, the Nielsen Corporation is a global marketing research firm. The company was founded in 1923 in Chicago, by Arthur C. Nielsen, Sr., in order to give marketers reliable and objective information on the impact of marketing and sales programs. One of Nielsen's best known creations is the Nielsen ratings, an audience measurement system that measures television, radio, and newspaper audiences in their respective media markets. Nielsen now studies consumers in more than 100 countries to provide a view of trends and habits worldwide and offers insights to help drive profitable growth.

- **Cognizant:** Starting as a joint venture between Dun & Bradstreet and Satyam Computers, the entity was originally designed to be the in-house IT operation for D&B. As the entity matured, it began to provide similar services outside of D&B. The entity was renamed Cognizant Technology Solutions to focus on IT services, while the former parent company of Cognizant Corporation was split into two companies: IMS Health and Nielsen Media Research. Cognizant Technology Solutions became a public subsidiary of IMS Health and was later spun off as a separate company. The fascinating aspect of this story is the amount of intellectual property, data, and capability that existed in this one relatively small part of Dun & Bradstreet. The interplay of data, along with technology services, formed the value proposition for the company.

- **IMS Health:** IMS became an independent company in 1998. IMS's competitive advantage comes from the network of drug manufacturers, wholesalers, retailers, pharmacies, hospitals, managed care providers, long-term care facilities and other facilities that it has developed over time. With more than 29,000 sources of data across that network, IMS has amassed tremendous data assets that are valuable to a number of constituents — pharmaceutical companies, researchers, and regulatory agencies, to name a few. Like Lewis Tappan's original company back in 1841, IMS recognized the value of a network of information that could be collected and then provided to others. In 2000, with over 10,000 data reports available, IMS introduced an online store, offering market intelligence for small pharmaceutical companies and businesses, enabling anyone with a credit card to access and download data for their productive use. This online store went a long way towards democratizing access to the data that had previously been primarily available to large enterprise buyers.

- **Moody's:** Moody's began in 1900, with the publishing of *Moody's Manual of Industrial and Miscellaneous Securities*. This manual provided in-depth statistics and information on stocks and bonds, and it quickly sold out. Through the ups and downs of a tumultuous period, Moody's ultimately decided to provide analysis, as opposed to just data. John Moody, the founder, believed that analysis of security values is what investors really wanted, as opposed to just raw data. This analysis of securities eventually evolved into a variety of services Moody's provides, including bond ratings, credit risk, research tools, related analysis, and ultimately data.

Dun & Bradstreet is perhaps the original innovator of the data-is-the-product business model. For many years, their reach and access to data was unsurpassed, creating an effective moat for competitive differentiation. However, as is often the case, the focus on narrow industries (like healthcare) and new methods for acquiring data have slowly brought a new class of competitors to the forefront.

COSTAR

Despite its lack of broad awareness, CoStar is a NASDAQ publicly traded company with revenues of $440 million, 2,500 employees, a stock price that has appreciated 204 percent over the last three years, a customer base that is unrivaled, and a treasure trove of data. CoStar's network and tools have enabled it to amass data on 4.2 million commercial real estate properties around the world. Simon Law, the Director of Research at CoStar, says, "We're number one for one very simple reason, and it's our research. No one else can do what we do." Here are some key metrics:

- 5.1 million data changes per day
- 10,000 calls per day to brokers and developers
- 500,000 properties canvased nationwide annually
- 1 million property photographs taken annually

CoStar has an abundance of riches when it comes to real estate data. Founded in 1987 by Andrew Florance, CoStar invested years becoming the leading provider of data about space available for lease, comparable sales information, tenant information, and many other factors. The data covers all commercial property types, ranging from office to multi-family to industrial to retail properties.

The company offers a set of subscription-based services, including

- **CoStar Property Professional:** The company's flagship product, which offers data on inventory of office, industrial, retail, and other commercial properties. It is used by commercial real estate professionals and others to analyze properties, market trends, and key factors that could impact food service or even construction.
- **CoStar Comps Professional:** Provides comparable sales information for nearly one million sales transactions primarily for the United States and the United Kingdom. This service includes deeds of trust for properties, along with space surveys and demographic information.
- **CoStar Tenant:** A prospecting and analytical tool utilized by professionals. The data profiles tenants, lease expirations, occupancy levels, and related information. It can be an effective business development tool for professionals looking to attract new tenants.
- **CoStarGo:** A mobile (iPad) application, merging the capabilities of Property Professional, Comps Professional, Tenant, and other data sources.

The value of these services are obviously determined by the quantity and quality of data. Accordingly, ensuring that the data remains relevant is a critical part of CoStar's business development strategy.

Since its inception, CoStar has grown organically, but also has accelerated growth through a series of strategic acquisitions. In 2012, CoStar acquired LoopNet, which is an online marketplace for the rental and sale of properties. CoStar's interest in the acquisition was less about the business (the marketplace) and much more about the data. Said another way, their acquisition strategy is about acquiring data assets, not people or technology assets (although those are often present). As a result of the acquisition, it is projected that CoStar will double their paid subscriber base to at 160,000 professionals, which represents about 15 percent of the approximately 1 million real estate professionals. Even more recently, in 2014, CoStar acquired Apartments.com, a digital alternative to classified ads. The war chest of data assets continues to grow.

The year 2008 was one of the most significant financial crises the world has seen. Financial institutions collapsed, and the real estate market entered a depression based on the hangover from subprime debt. Certainly, you would expect a company such as CoStar to see a similar collapse, given their dependence on the real estate market. But that's not exactly what happened.

From 2008 to 2009, CoStar saw an insignificant revenue drop of about 1 percent. This drop was followed by an exponential rebound to growth in 2010 and beyond. Is it possible that data assets create a recession-proof business model?

	2008	2009	2010	2011	2012	2013 (9)
Revenues	$ 212,428	$ 209,659	$ 226,260	$ 251,738	$ 349,936	$ 310,762
Net Income	$ 24,623	$ 18,693	$ 13,289	$ 14,656	$ 9,915	$ 16,935
EBITDA	$ 56,589	$ 46,511	$ 36,474	$ 34,623	$ 60,192	$ 62,892

CoStar Financial Results

While there are other data alternatives in the market (Reis Reports, Xceligent, CompStak, ProspectNow, and others), the largest collection of data is a differentiator. In fact, it is a defensible moat that makes it very hard for any other competitors to enter. For CoStar, data is the product, the business, and a control point in the industry.

IHS

In 1959, Richard O'Brien founded IHS, a provider of product catalog databases on microfilm for aerospace engineers. O'Brien, an engineer himself, saw how difficult it was to design products and knew that utilizing common components could dramatically increase the productivity of engineers. However, he took it one step further by applying technology to the problem — using electronic databases and CD-ROMs to deliver the knowledge furthered the productivity gains. And engineers love productivity gains.

This attitude toward data was set in the company's DNA from the start as the company could see how to improve the lives and jobs of their clients, just through a better application of data. In the 1990s, IHS started distributing their data over the Internet, and with an even more cost-effective way to share data assets, they decided to expand into other industries. Soon enough, IHS had a presence in the automotive industry, construction, and electronics.

As seen with CoStar, IHS quickly realized that business development for a data business could be accelerated through acquisitions. Between 2010 and 2013, they acquired 31 companies. This acquisition tear continued, with the recent high-profile $1.4-billion acquisition of R.L. Polk & Company. As the parent company of CARFAX, R.L. Polk cemented IHS's relevance in the automotive industry.

IHS's stated strategy is to leverage their data assets across interconnected supply chains of a variety of industries. For their focus industries, there is $32 trillion of annual spending in those companies. IHS data assets can be utilized to enhance, eliminate, or streamline that spending, which makes them an indispensible part of the supply chain. IHS's data expertise lies in

- Renewable energy
- Chemicals
- Technology
- Automotive parts
- Aerospace and defense
- Maritime logistics

IHS also has a broad mix of data from different disciplines.

While some view data as a competitive differentiator or something to augment current offerings, CoStar, IHS, and D&B are examples of companies that have a much broader view of data: a highly profitable and defensible business model.

MEATPACKING DISTRICT (CONTINUED)

The role of data in enterprises has evolved over time. Initially, data was used for competitive advantage to support a business model. This evolved to data being used to augment or improve existing products and services. Both of these applications of data are relevant historically and in the future. However, companies leading in the data era are quickly shifting to a business model of data as the product. While there have been examples of this in history as discussed in this chapter, we are at the dawn of a new era, the Data era, where this approach will become mainstream.

The company that started as a social bookmarking service quickly realized the value of that data that they were able to collect via the service. This allowed them to build a product strategy around the data they collected, instead of around the service that they offered. This opportunity is available to many businesses, if they choose to act on it.

PATTERNS IN BIG DATA

THE LEADERS OF the data revolution will be able to recognize patterns, understand why they are present, analyze them, and apply them to create value.

Each story in Part I of this book reveals a data factor. *Data factors* are the primary drivers of the value that exists as a result of improved consumption and application of data. The data factor itself may be a pattern. But in most cases, the data factor exposes a broader set of patterns. The patterns become something that can be acted on in any business where it is relevant.

This chapter identifies 30 data factors, pulled directly from the stories in Part I, and in each case, they represent a potential initiative and opportunity for those that want to lead the revolution.

The data factors are then further deconstructed into a set of big data patterns: 54, to be precise. These patterns represent the best practices, approaches, and considerations for each of the data factors.

THE DATA FACTOR

Before delving into and extracting the critical patterns in big data from the stories in Part I, it's useful to quickly recap the stories covered so far and the key data factors behind each story:

- **Farming and agriculture data factors (Chapter 1):**
  - Creating and utilizing new sources of data (drones, geospatial data collection)
  - Building new data applications (FieldScripts, CanopyCheck)
  - Transforming and creating new business processes (integrated farming, predictive maintenance)
- **Medical data factors (Chapter 2):**
  - Frequent monitoring and data collection (Quanttus)
  - Redefining a skilled worker (eliminating generalist practitioners, evolving the skills of physicians)
  - Exposing opinion-based biases (Ben Goldacre)
  - Social networks leveraging and creating data (HealthTap)
  - Deconstructing the traditional healthcare value chain
- **Insurance data factors (Chapter 3):**
  - Transforming business processes (from actuarial science to data science)
  - Real-time monitoring and decision making (dynamic risk management)
  - New product offerings (pay-as-you-drive)
  - Risk modeling (RMS, AIR, EQECAT)
- **Retail and fashion data factors (Chapter 4):**
  - Defining new channels to market (Stitch Fix, Keaton Row)
  - New economic models (Zara, Amazon)
  - Building for customers instead of markets (personalization)
- **Customer Service data factors (Chapter 5):**
  - Changing the definition of a product (Boeing)
  - Democratizing access to data (automobile manufacturer)
  - Creating meaningful relationships through data (Zendesk)
- **Intelligent machines data factors (Chapter 6):**
  - New economic models (Vestas)
  - Data collection for competitive advantage (drones)
  - Changing the definition of a product (Tesla)

- **Government and society data factors (Chapter 7):**
  - Changing incentives and perception
  - Finding a balance between privacy and insight
  - Data security
  - New partnerships on the foundation of data
- **Corporate sustainability data factors (Chapter 8):**
  - Real-time monitoring and decision making (early warning systems)
  - The impact on the environmental footprint
  - Building resiliency and sustainability
- **Weather and energy data factors (Chapter 9):**
  - Forecasting and predicting future events
  - Effect on renewable energy and consumption

By reviewing the data factor present in each of these stories, patterns begin to emerge. These patterns can be acted on now, for those inclined to lead the data era.

SUMMARY OF BIG DATA PATTERNS

This book discusses 9 stories and 30 data factors. Across those stories and data factors, a number of patterns begin to emerge. The 54 big data patterns described here represent the patterns that can be applied in virtually any organization today.

REDEFINING A SKILLED WORKER

The Data era is demanding a new definition of skilled workers. While this may require skills like statistics or math, that is merely one aspect of the skill gap that must be filled. In medicine, it's about redefining medical school to include skills like data analysis and data collection. In farming, the new skill set involves understanding how to utilize multiple sources of data (from drones, GPS, or otherwise) and apply that insight to deliver better yields and productivity.

The definition of a skilled worker has changed dramatically, based on different eras. In the 1700s and 1800s, skilled workers were defined by either their physical abilities or their knowledge of a certain craft (think book-keeper). At the time of the Industrial Revolution, skilled workers were defined by their physical capacity to operate a machine or work on an assembly line. In the last 20 years, skilled workers have evolved further, with a premium placed on customer service and technology skills.

A skilled worker in the next decade will be defined by his ability to acquire, analyze, and utilize data and information. This new skilled worker will emerge in every industry, with a slightly different definition of the skill set for each industry.

The key patterns to redefine a skilled worker are

1. Understanding the skill sets needed today, tomorrow, and further in the future, based on the potential for data disruption.
2. Redefining roles and skill sets to take advantage of the new data available that can impact business processes.
3. Training and retraining current and new workers is a distinguishing capability to remain relevant.

CREATING AND UTILIZING NEW SOURCES OF DATA

With the rise of intelligent machines and more economical means of collecting and analyzing data, harnessing new sources of data will become a differentiating capability across a wide range of industries.

In Chapter 1, it was evident how new sources of data are transforming farming and agriculture. Twenty years ago, a farmer would have never thought about how to utilize satellite images of her property, let alone have the ability to generate this data on a frequent basis. This phenomena is emerging in every industry.

In healthcare, data sources have evolved from medical records created a few times a year to continuous data creation through the usage of wearables and mobile health solutions. That stream of data can then be combined with social data to create a new and richer fusion of data.

The key patterns for new sources of data are

4. Creating new sources of data may lead to competitive advantage. Most companies have a current source of data and do not think about external sources where they can obtain new data.
5. Joining previously disconnected or disparate datasets can lead to new insights and competitive advantages.
6. New data types emerge from the fusion of internal and new external sources of data.

BUILDING NEW DATA APPLICATIONS

Data is valuable only if it's useable and consumable. Data applications are a means for making data accessible. Even with a new generation and definition of skilled workers, it will not be acceptable to give the average worker a giant dataset, expecting him to comb through it for insight and action. Instead, data applications will become the vehicle for delivering and utilizing data.

CanopyCheck, the farming application introduced in Chapter 1, is an example of a data application. Even with the appropriate training, the average potato farmer should not be expected to be able to collect and analyze groundcover data by herself. CanopyCheck fills that gap. The burden is still on the farmer to ensure adequate and reliable data collection (part of the new skill set needed), but CanopyCheck provides a simplified way to analyze, visualize, and then act on the data.

With the right set of data applications, the scope of the training challenge is reduced dramatically. Instead of having to redefine a skilled worker as a data scientist, the data application can enable a skilled worker to work like a data scientist (without actually being one).

The key patterns in building new data applications are

7. New data applications will enable faster and more productive use of data.
8. Data applications will evolve as datasets evolve, ensuring continuous insight and learning.

TRANSFORMING AND CREATING NEW BUSINESS PROCESSES

A typical fashion retailer will start their buying process for a fall collection in the winter prior. They will review designs, select suppliers, place orders, and then roll them out into stores. Once the designs are in stores, the retailer assesses what is selling and utilizes that information to place additional orders. These are the business processes that drive the fashion retail industry.

In Chapter 4, StitchFix demonstrated how they are creating an entirely new type of business by both transforming and creating new businesses processes based on data. StitchFix personalizes a fix (or personalized recommendation) for each of their clients and incorporates that feedback into their buying and production processes, thereby creating a new process for identifying what to source and build.

In Chapter 3, the introduction of dynamic risk management is a fundamentally new approach and business process for the insurance industry. Historically, every process in an insurance company is based on an actuarial assessment of the individual requesting insurance, followed by an offer and premium. In the era of dynamic risk management, these decisions are made as the individual consumes the insurance product (i.e. pay-as-you-drive insurance). Dynamic insurance explicitly connects the risk and premium and creates a feedback loop that can benefit both insurer and customer.

The key patterns in transforming business processes are

9. Data will transform current business processes and introduce new ones.

10. Data enables the personalization of business processes to an individual level instead of customer segments.

11. Companies are being born solely on the basis of harnessing data, disrupting incumbent companies that use traditional business processes and approaches.

DATA COLLECTION FOR COMPETITIVE ADVANTAGE

It's very easy to collect data from a cash register. Every sales transaction is entered and can be easily loaded into a database for analysis. The collection is easy and typically is performed automatically. It is much harder to collect data from a drone that docks only to recharge. Plus, a drone can collect data in the form of video, sensor readings, or nearly anything it is tasked to do. So, which do you choose?

It is also difficult to collect data from a remote oil rig, much like it is hard to pull data from a tractor that may have little or no connectivity to a network in the field in which it is used.

Data collection can happen in a number of places. It can come from a device on the edge of a network, an individual close to the data source, a connected machine, or somewhere else. While the capability to collect data is often taken for granted (or, in some cases, deemed impossible), the Data era will place an increased emphasis on the ability to collect and synthesize data. As big data starts to deliver competitive advantages, the technology for collecting data will become more widespread.

The key patterns in data collection are

12. Collecting and synthesizing data will be a core competence in the Data era.

13. Data that is collected must be transmitted and analyzed in a timely fashion.

14. Data must be collected over the lifecycle of a company, product, or business process. Data at a point in time is often interesting but not necessarily informative. Gathering data is one thing, but ensuring that it is correct and actionable is another.

EXPOSING OPINION-BASED BIASES

Every company, organization, and industry has a set of what are deemed to be unchallengeable truths. Whether it is because the leadership team has seen the patterns through the years, or there is simply a lack of organization fortitude; the unchallengeable truths exist.

A sacred cow in the Data era can be likened to the opinion-based biases discussed in Chapter 2. Recall that a study by the Cleveland Clinic found that doctors disagree on initial diagnoses 11 percent of the time, and in 18 percent of cases, doctors recommend major changes to previously prescribed treatment. This is what happens when decisions are based on opinion.

Does the Data era eliminate the need for opinions? Of course not. Insight, intellect, and opinion are critical for rigorous decision making. But opinions must be based on data and hard evidence. Sacred cows, by their very nature, encourage an organization to start with opinion, history, or gut feeling instead of data.

The key patterns to expose opinion-based biases are

15. An individual or organization must document their biases (or sacred cows) up front to ensure those biases are recognized and understood.

16. Decisions will start with data, not opinion or organizational sacred cows. To ensure this is happening, force every decision to be supported by data and carefully documented for future analysis. If it cannot be, then you can explore why (i.e. was the data uncollectable, unavailable, unable to be analyzed?).

REAL-TIME MONITORING AND DECISION MAKING

For decades, the default way to analyze data has been based on collecting data, loading it into a system, then analyzing the data. It's a simplistic process, but something that has been proven to provide insights.

But what happens if that is not sufficient? What happens if decision making is dependent on analyzing a constant flow of information and making decisions in the moment? If that is the requirement, then the traditional collect-load-analyze process is not sufficient.

In 2013, the NeoNatal Intensive Care unit at the Hospital for Sick Children in Toronto announced that it would start real-time monitoring of babies. The hospital had discovered that an infection called late-onset neonatal sepsis (a blood infection) typically occurs between days 8 and 89 in an infant's life.

A traditional approach of collecting, loading, and analyzing data would not be sufficient, as discovering this a few days, weeks, or months after the fact would be too late. Therefore, the Toronto hospital turned to real-time monitoring for real-time decision making to transform the treatment of infants. The flow of real-time data, in this situation, could be the difference between life and death.

The key patterns in real-time monitoring are

17. Some data is time sensitive. Time-sensitive data must be collected in a continuous manner, producing a data stream.

18. Time-sensitive data must be analyzed as it generated using data-stream analytics so it can be acted upon immediately.

SOCIAL NETWORKS LEVERAGING AND CREATING DATA

Networks of individuals, organizations, or machines can be powerful in the Data era. Without a network of data, the dataset is, by definition, limited. Limited data leads to decisions being more opinion based, which can lead to less constructive decisions.

In Chapter 2, HealthTap demonstrated how the social network of patient data could lead to better outcomes and decisions for the rest of the network. In HealthTap's business model, the social aspect of the data enables self-service by the individuals. Self-service removes the need to involve a physician in every decision, which speeds up the process to reach a conclusion.

James Surowiecki wrote *The Wisdom of Crowds* (Anchor) in 2004, on the cusp of the Data era. His primary assertion was that large groups often make better decisions than any single member of the group. This is due in part to the fact that the diversity of independently deciding individuals, when analyzed as a group, provides a broader perspective than an individual ever could. This is precisely why social-network data can be so powerful. It's not because any one individual is right (although they may be), it's because the broad data that can be collected provides critical insight.

In summary, the key patterns in social networks are

19. Data from social networks provide a diverse set of opinions, enabling more exhaustive analysis and often better outcomes.

20. Data from social networks, cultivated from crowd sentiment, will often be more accurate than the opinion of a single expert.

DECONSTRUCTING THE VALUE CHAIN

Over time, every industry develops a *value chain:* the set of processes and events that create value for each stakeholder. For example, in the automotive industry, the traditional value chain includes parts suppliers (value created from developing high-quality parts), automotive manufacturers (value created from assembling parts and overall design and performance), automotive dealers (value created by providing sales and marketing, and access to inventory), and the consumer. Data has the ability to deconstruct the value chain by either removing parts of the chain or redefining parts.

In the case of Tesla, which was covered in Chapter 6, data has removed parts of the traditional value chain. With the intelligence provided by Tesla cars, there is little need for automotive dealers. Hence, Tesla has built a business model that excludes traditional dealer networks, instead opting to build their own company-owned dealerships. Tesla is now taking the deconstruction a step further with their initiative to build Gigafactories to produce batteries for their cars. This step will start to deconstruct the parts suppliers portion of the value chain.

The deconstruction of the value chain is possible because Tesla has chosen to build intelligent machines instead of just automobiles. The data generated at each step enables the company to optimize sales, support, and service, thereby making it economical for a single company to be vertically integrated along the value chain.

The key patterns in deconstructing the value chain are

21. Utilize data to remove steps in the value chain and improve the end product.

22. Understand how data can transform, expedite, or reduce the cost of steps in the value chain.

NEW PRODUCT OFFERINGS

Data can act as a differentiator, transforming business processes or eliminating steps in the value chain. In most cases, companies view data as a means to an end, utilizing the assets to augment current offerings or improve traditional processes. However, the Data era brings a much broader opportunity: Data can become the product.

Companies like Nielsen, IMS Health, and others have known this for decades. Their core businesses are built on the notion that data is the critical asset, and they have business models built to monetize data. This opportunity is available to nearly every company, if they can exploit their data without eroding their existing competitive advantage.

Monsanto has extended their business by allowing data to become a product. The data that they collect on farming and the effectiveness of different techniques offers tremendous value to farmers. Accordingly, they monetize the data itself, as well as the products that the data suggests are required to improve crop yields.

The key patterns for new product offerings are

23. Competitive advantage can be gained by leveraging data or by monetizing data assets.
24. Data must be consumable by external parties if there is a goal of monetization.
25. Companies should not monetize data that erodes their existing competitive advantage.

BUILDING FOR CUSTOMERS INSTEAD OF MARKETS

The era of market segmentation, focus groups, and mass mailings should be coming to an end. The only thing slowing this evolution is the lack of understanding of how data can change the status quo, the foresight to understand how to do it, and the tools to make it happen. But it's coming.

Marketing and selling will become a much more personalized initiative, with the winner being the fastest to achieve personalization. The economics are fundamentally different when a company is marketing and selling to a customer of one.

The best examples of this were highlighted in Chapter 4. Keaton Row is founded on the principle of a personal touch, with all sales done in a personalized manner. Stitch Fix, after collecting data for a short period of time, becomes a truly personalized and unique experience. Both are accomplished through the use of data about customer behavior and individual preferences.

The key patterns in building for customers, instead of markets, are

26. Data will drive the personalization of sales and marketing.
27. With personalization, data can change the economics of many industries and business processes.

TRADEOFF BETWEEN PRIVACY AND INSIGHT

While personalization will be powerful for companies to transform marketing and sales, it can have the impact of affecting the privacy of individuals. The perception of many individuals may shift to the feeling that their activity is being tracked or monitored. This presents a classic tradeoff to a consumer: Are you willing to sacrifice some privacy for the rewards of personalization?

Companies utilizing data assets for personalization need to be cognizant of this tradeoff and ensure that policies allow for consumers to be engaged in the manner that they choose.

The key patterns are

28. Privacy and personalization are often at odds with each other.
29. Understanding how to utilize personalization, without overtly sacrificing privacy, is critical to appropriate data stewardship.

CHANGING THE DEFINITION OF A PRODUCT

William Davidow wrote *Marketing High Technology* (Free Press) in 1986. While we have seen many phases in technology since then, it's a timeless piece of work on how to think about building great products. The fundamental message of the book is that a product-development organization has to think about building a complete product, not just a product. A *product* is something that a client can buy. But a *complete product* addresses all the things around and related to the product: market fit, distribution channels, sales, service, marketing, and positioning. Data changes the notion of a complete product, either by augmenting current offerings with insight or by enabling data to be the product itself.

Davidow's wisdom includes things like

- Marketing must invent complete products and drive them to commanding positions in defensible market segments.
- The cost of creating a complete product is often many times the cost of developing the product.
- Serviceability must be designed into a product.
- Great products need a soul.
- Companies fail because they are incapable of delivering total customer satisfaction.

Davidow's work, updated for the Data era, would offer a compelling definition of a complete product. The Tesla example in Chapter 6 illustrates how the product is much more than just a car (for example, after-sale feature additions), thanks to the innovative use of data.

The key patterns in changing the definition of a product are

30. With data, the definition of a product may extend beyond features and function to experience and serviceability.

31. Data can change products from being transactional in nature to iconic and desirable, offering long-term relationships with customers.

INVERTING THE SEARCH PARADIGM FOR DATA DISCOVERY

As data volume grows and organizations begin to derive value from this data, the request from the organization is what would be expected: more access to more data as soon as possible. But it is not necessarily easy to make data available to an organization, as data is not easily consumable and often has to respect certain guidelines or policies.

The traditional method of finding information is text-based search. A search is the default approach because it is something that everyone is familiar with, given the role that Google and other online search engines play in providing data access. The problem with searching is that it requires the user to know what she is looking for. And, as we learned in the stories in Part I, many of the revolutions with data come from data that the users would never think to ask for in the first place.

To effectively democratize data access, an organization must consciously decide to make it available (with the right processes and procedures) and serve the data up to the right users at the right time. This is the opposite of simply providing a search capability, and accordingly, is not something found in most organizations today.

The key patterns in data discovery are

32. The democratization of data is starting to happen, putting an impetus on companies to organize and govern the data made available within the company.

33. A data search answers only questions that the user knows to ask. So effective access to data must serve up the right data to the right person at the right time, without the user having to request it.

DATA SECURITY

As data is democratized, a premium is placed on data security and data governance. Given the belief that data is a core asset to many companies (and, in some cases, *is* the company), data security and governance is arguably as important as leveraging data in the first place.

Data security is about setting the parameters for who can access what and when. Data security treats data as a corporate asset, using the same rigor that a company would to protect physical assets. As data relates to clients and customers, security issues may represent a serious reputational risk with long-term consequences.

Data governance is the approach to managing the lifecycle of data and maintaining history of who has accessed what data and when. It's impossible to have a data security strategy without incorporating governance and vice versa.

The key patterns in data security are

34. Security of data is just as important as leveraging data for insights and business transformation.

35. Data governance is necessary to manage the lifecycle of data assets to ensure data security and compliance with business or legal practices.

NEW PARTNERSHIPS FOUNDED ON DATA

IBM and Apple: No one could have predicted that partnership. Steve Jobs once famously stood in front of an IBM building and was photographed making an obscene gesture towards the IBM logo. It was an emotional reaction to what he viewed as his arch nemesis at the time. However, time — and, in this case, data — can create the unlikeliest of partnerships.

The foundation of IBM and Apple partnering in 2014 is data; more specifically, enterprise data. Apple brings devices, users, and an affinity for design and user engagement. IBM brings rich enterprise client relationships and access to data. Together, the partnership will unlock enterprise productivity by exploiting data. This is just one example of new partnerships based on data that will emerge.

New partnerships will form on the idea that combining previously disparate datasets can lead to new insights, new customers, or new markets. This is not a natural consideration for many companies, so it will require a few determined individuals to identify and lead previously unconsidered partnership ideas.

The key patterns are

36. New partnerships will emerge as a result of innovation driven primarily by data.

37. New insights arise from combining the assets of separate companies, starting with their data assets.

SHORTENING THE INNOVATION LIFECYCLE

Stitch Fix and Zara have both disrupted retail by shrinking the innovation lifecycle in retail. A typical brick-and-mortar retailer requires 9 to 12 months to understand changing consumer preference and to then drive that through their innovation lifecycle (buyers-manufacturers-materials) and back to stores. Zara does it in two weeks based on the collection and application of data. Stitch Fix does it in the moment, as the preferences discovered in one fix will be reflected in the next one to ship. The innovation lifecycle is compressed.

The compression of innovation lifecycles will occur in nearly every industry. The leaders will do it now, while the followers will wait until they have been sufficiently disrupted as to have no other option but to react. Applying many of the patterns discussed in this chapter can shorten innovation lifecycles.

In summary, the key patterns are

38. Understand where the lack of data is extending the innovation lifecycle and eliminate the constraint.

39. Re-examine the innovation lifecycle to automate or remove parts of the cycle with the application of data.

DEFINING NEW CHANNELS TO MARKET

Keaton Row is forging a revolution in fashion. While most retailers innovate through catalogs, storefronts, or websites, Keaton Row has defined an entirely new channel to market: women with computers, coupled with a keen sense for fashion.

This channel to market capitalizes on one of the major factors that women use in buying clothes: recommendations from a person whom they trust. Keaton Row's business model would not be economical if it required stores or networks of friends to join. Instead, it relies on data to match stylists with consumers. Then, it gives stylists the data and information needed to make recommendations as a trusted source to their clients.

The key patterns in defining new channels to market are

40. New channels to market can change the economics of an industry.

41. Data can unlock new channels to customers.

NEW ECONOMIC MODELS

General Electric famously changed the economics for aircraft engines through price innovation. Instead of charging simply for the engines themselves, they shifted to charge for engine uptime.

> *A master application of pricing innovation was General Electric's reinvention of the commercial aircraft engine business. Engine manufacturers had traditionally used engines as a loss leader to secure the lucrative business of selling replacement parts. When several trends threatened that model, GE offered airlines the option of buying power by the hour — in essence, purchasing as a package engines, parts and MRO services (maintenance, repair and overhaul) and paying on the basis of uptime, or per hour of use. The strategy has allowed the company to make the most of its competitive advantages (its technical skills, diagnostic capabilities, financing expertise and scale, and end-to-end MRO services) against major competitors.*
>
> The Boston Consulting Group (*http://marketing-stg.bcg.com/content/ articles/pricing_consumer_insight_pricing_myopia/*)

This approach aligned GE's client interests with its own. This innovation was possible because of data. Specifically, it was engine and usage data that GE could collect to effectively bill their clients in this manner. Data will continue to create new economic models.

Vestas, the wind turbine maker, is pursuing similar innovation in their economic models. Clients of Vestas have typically only outlaid cash upfront for the large capital investment of a turbine. This payment process creates difficulty for the client and a one-time revenue event for Vestas — not optimal for either party.

However, with the collection and application of data, Vestas can start to monetize insight over the life of the turbine. This may include information about how to operate the turbine for optimal productive use or could be as simple as predictive maintenance. In either case, data enables those new economic models.

As an example of new economic models, consider how the insurance sector will be disrupted by the entry of mobile network operators into this space. Insurers tend to operate in a static, slowly evolving world where historical risks are used as a proxy for future risks and financial premiums are collected annually. On the other hand, mobile network operators have the ability to monitor usage in real-time and offer multiple payment options. In addition, the flow of data and information via sensors and mobile technology will enable a whole new class of dynamic insurance products.

The key patterns of new economic models are

42. Data will create new economic models, often providing the ability to better align customer and supplier interests.
43. New pricing or economic models based on data can disrupt incumbent and entrenched suppliers.

FORECASTING AND PREDICTING FUTURE EVENTS

Most industries would like to have access to a crystal ball in order to know what opportunities and challenges lay around the corner. Despite the fact that it is difficult to forecast complex systems, big data is increasingly providing sufficient information (relating to human activity, as well as our environment) to identify patterns and improve the ability to forecast events that can help to both maximize rewards and minimize risks.

The key patterns are

44. Data provides the basis for predicting future events, outcomes, and impacts.
45. Forecasting and predicting is not valuable unless it is acted upon. Data often holds the answer to what action should be taken.
46. Predictive analytics requires the fusion of knowledge and both historical and real-time data, obtained from internal and external sources.

CHANGING INCENTIVES

Success in business tends to follow the formula of knowing what you can do and doing it well. Without disruption, this formula is probably a safe one to follow. The big data revolution will create new winners, and these organizations will be capable of utilizing data to develop strategies. For this reason, workers should be encouraged to move out of their organizational silos, which generally means moving them out of their comfort zones, in order to consider the implications of big data for their part of value creation in the business. Incentives are needed to facilitate sufficient amounts of collaboration around sharing data. Furthermore, incentives and bonuses are needed to compensate individuals for personal risks when considering the rewards for the business as a whole.

The key patterns in changing incentives are

47. Incentives will frequently determine the willingness of an organization or individual to take risk.

48. A proper incentive framework may determine success (or failure) in an initiative to apply data to a business problem.

49. Organizational boundaries or silos lead to suboptimal outcomes and insights. Incentives must be used to break down this barrier.

NEW PARTNERSHIPS (PUBLIC/PRIVATE)

The challenges of addressing some global trends are so immense that business and government are realizing that the only viable solution is to form public-private partnerships. Regarding big data, this is important as government could play a central role in providing the IT infrastructure for collecting and distributing data that has a common societal good.

The key patterns are

50. Cooperation between the public and private sector in creating innovation from big data.

51. Construction of adequate IT infrastructure for big data.

REAL-TIME MONITORING AND DECISION MAKING (EARLY WARNING SYSTEMS)

Both business and government are improving their ability to build a risk register and then setting up a process to monitor and respond to these risks. By automating the process, it will be possible to detect early signs of a growing number of issues and to react by making appropriate decisions.

The key patterns are

52. Novel sources of data can help to monitor and respond to risks.

53. Construct early warning systems for the most dangerous risks.

54. Be prepared and have action plans in place to allow rapid responses to early warning systems.

A FRAMEWORK FOR BIG DATA PATTERNS

This chapter has identified 54 patterns of big data, synthesized from a variety of industries, stories, and companies. As mentioned in Chapter 10, pattern recognition can be a gift or a trap. Patterns are a gift when extracted from success stories and when they can be applied as part of a framework or methodology for assessing market opportunities.

Part III of this book is about leading the revolution. It will explore the opportunity that exists in the data revolution, suggest a methodology for assessing and exploiting that opportunity, and finally recommend an architecture for capitalizing the opportunity.

LEADING THE REVOLUTION

THE DATA OPPORTUNITY

WHAT OIL TEACHES US ABOUT DATA

NINETY PERCENT OF the data in the world has been created in the past two years. Eighty percent of that data is unstructured in form. Data offers a business opportunity, assuming that an organization can tap into that resource and put it to productive use. This requires an organization to identify quality sources of data and learn how to refine that data for the specific needs of the company. In that way, the process for finding and refining oil can teach us a lot about how to exploit the new natural resource in enterprises: data.

Manufacturing oil is a multistage process, requiring a variety of different tools and considerations at each step. Each step in the process is critical to the next step, otherwise the quality of the end product will be diminished or, worst case, unusable. The seven major steps are

> 1. **Exploration:** Without finding oil sands, or another source of crude, and hence petroleum, the process can never begin. Petroleum is most often found in areas of porous rock. Once petroleum has been located, it's prudent to do some test drilling to ensure that more drilling will be productive.

2. **Drilling:** While exploration requires tools such as geospatial maps and visualizations, drilling requires a rotary drill and insertion of a pipe. Typically, petroleum will be thousands of feet deep into the rock. Accordingly, a drill must be able to penetrate deeply.

3. **Recovery:** Pipes and valves are installed into the rock to ensure the proper extraction of the crude oil. These pipes are connected to a recovery system, which separates the oil and gas.

4. **Distillation:** Fractional distillation is the process of removing gasoline from other compounds in the crude oil. It's the process of heating crude oil, letting it vaporize, and then condensing the vapor, prior to the refining process.

5. **Refining:** This is one of the most important processes in the manufacturing of oil. Catalysts are added to the crude oil to ensure that the oil is useable. This is where the quality of the gas is largely determined and requires a precise approach.

6. **Additives:** Chemicals are added to ensure performance of the finished product. This will help extend the life of gas engines and ensure optimal operation.

7. **Transporting:** The last step in the process is about ensuring that the finished product is delivered to a location where it can be consumed.

Organizations that have the ability to tap into data as their primary natural resource will drive competitive advantage versus their peer group. That is the data opportunity. Leading organizations will exploit the opportunity, drawing parallels from the oil refining steps:

1. **Exploration:** Data can be anywhere in the organization or available from external sources (some in the public domain and others from fee-based services). Leaders in the Data era will explore beyond the obvious places, looking for new data assets which can be utilized for competitive advantage.

2. **Drilling:** Once data is located, the right tools must be in place to access the data. But, it's more than just tools; it's the approach, the methodology, and the techniques to ensure that the effort will yield results.

3. **Recovery:** Once data is located and accessed, there must be a strategy to recover and extract the data. From that point, the data can be moved to the right part of the organization.

4. **Distillation:** Just like consuming tainted oil will prevent an automobile from running, consuming poor-quality data can cripple an organization. The data must be distilled, detecting and fixing errors and missing values, before it is refined, to ensure that it can be utilized.

5. **Refining:** This is where the data is turned into action. It's evolved into a state where it can be consumed across the organization, by people, applications, or business processes.

6. **Additives:** The data can be enriched, with third-party data or simply through human aided curation. This enhances the value of the asset.

7. **Transporting:** Data must be in the right place, at the moment of a decision, in order to impact an organization. This puts a premium on the transportation and application of data, across an organization.

BAIN STUDY

In a 2013 Bain & Company report titled "Big Data: The organizational challenge," Travis Pearson and Rasmus Wegener assert that early adopters of big data are gaining a substantial lead over the rest of their peers. Their study, which analyzed 400 large companies, found that the leaders are

- Twice as likely to be in the top quartile of financial performance within their industries
- Five times as likely to make decisions faster than market peers
- Three times as likely to execute decisions as intended
- Twice as likely to use data very frequently when making decisions

Even more striking is the difference between the top performers and the worst performers. From this analysis, data is much more than an opportunity; it's an imperative.

With the assumption that data is an imperative, the Bain study then begins to look at how an organization will exploit data for competitive advantage. The challenge is one of ownership. Should business units lead? Or a central IT group? Or a combination? The conclusion from Bain is that it depends on which phase of adoption the organization is in.

The Bain study describes three phases for data adoption:

- **Strategy Phase:** The leadership sets the data strategy, articulates the vision and objectives, and mobilizes a team to build out the strategy and prepare for execution. In this phase, leadership and accountability must be assigned, otherwise the organization will suffer from too many opinions and too little action.
- **Execution Phase:** The execution phase, similar to the oil refining process, is where the data assets are converted into useful assets to benefit the organization. This is where data is accessed, collected, and ultimately turned into a consumable form of knowledge and insight.

- **Infrastructure Phase:** The infrastructure phase is where the IT environment is evolved to support the initiatives of the data strategy. This may include the physical infrastructure, along with policies and procedures to govern the lifecycle of the data assets.

The stories in this book illustrate how the opportunity varies based on industry. In farming, the opportunity may come from productivity, through a better application of data. In medicine, it's moving from opinion-based medicine to data science. And in insurance, it's about having a full view of your customer, providing an immediate assessment of potential risk and reward, at the time of engagement. Those opportunities are available only if you have the data and a plan to seize the opportunity.

SEIZING THE OPPORTUNITY

Organizations around the globe are on the cusp of seizing the data opportunity. Like the best oil companies in the world, they will compete based on their ability to explore, drill, refine, and transport the raw material. While many companies are willing to explore the application of new software and IT systems when handed the task of seizing the data opportunity, the leaders of the Data era will start with the appropriate leadership team, planning, and long-term strategy.

With the appropriate strategy in place, the best-fit tools can then be applied to seize the opportunity.

14

PORSCHE

ROME

ROME IS UNIQUE on the world stage. The city, known for its ancient artifacts and traditional cuisine, also possesses the expanse and calling of the Vatican, neatly nestled along the Tiber river, which snakes through the city. While known for its major monuments and sights, I prefer the backstreets of Trastevere or the winding alleys of the Jewish ghetto. These are the places where local craftsmen are found, with the creations that define each of their life's work.

I was just outside the Jewish ghetto in Rome, when I happened upon a shop that sold only belts. This was striking in that often times belts are relegated to an accessory rack in a shoe store. Or belts may represent less than 1 percent of a clothing retail space. But never had I seen a shop dedicated to belts. Now, it was not a large store, but it was more than sufficient to tell the story of this craftsman owner and his passion.

In a brief discussion, mixing my poor Italian with his better English, I learned a bit of his history in making belts, which is something that he had been doing for over 50 years. As he showed me the detail on each one, I marveled at the attention to detail: the stitching, the lines, the widths, and the more obvious individualities of color and buckle.

The shop owner was much more interested in where I planned to wear the belt than which designs I liked. In his mind, the selection of a belt was much more about how I planned to use it, than which one I preferred on the shelf. In his partial English, I gleaned that he wanted to know "my purpose" for the belt.

The Macmillan Dictionary defines the concept of *fit for purpose* as something that "is good enough to do the job it was designed to do." Hence, the fit for purpose in the belt store was about ensuring that I chose a belt that could do the job required, at the place I intended to wear it.

FERDINAND PORSCHE

Ferdinand Porsche was born in 1875 in Maffersdorf (now the Czech Republic). He developed an early fascination with technology — electricity, in particular. His first job, at the age of 18, was at Bela Egger & Co., the first factory for electric lighting and power transfer in Austria.

The young Porsche quickly impressed his management, rising to positions of increasing responsibility. He distinguished himself through an inventor's mindset, constantly tinkering to make existing designs better and, at other times, creating his own. His first breakthrough came when he built an electric wheel hub motor, which he would soon have the opportunity to race at an event in Vienna. His design showed well, leaving an impression that would last for years to come.

Three years later, his improved electric wheel hub motor powered a non-transmission vehicle at the World's Fair in Paris. The vehicle, developed by Hofwagenfabrik Jacob Lohner & Co., was known for its innovative and sleek design. The vehicle that was entered in the World's Fair was known as Lohner-Porsche, a testament to the combined innovation present in the design.

Porsche went on to work at Lohner for eight years, developing his technology and design skills. Seeking a new challenge, he joined Austro-Daimler as a technical manager, quickly moving along to Daimler-Motoren-Gesellschaft in Stuttgart, the seminal home of Mercedes Benz. As a technical manager and innovator at Daimler, he oversaw the creation of the Mercedes compressor car. Porsche, through experience, was honing his ability to customize a design or feature, to appeal to a specific audience or task. This skill would be paramount in the formation and evolution of Porsche motor cars.

THE BIRTH OF PORSCHE

In 1931, chasing his growing passion for innovation, Ferdinand Porsche left Daimler to start his own company, which was named Dr. Ing. h. c. F. Porsche GmbH, Konstruktionen und Beratung für Motoren und Fahrzeuge. His son,

Ferdinand "Ferry" Anton Ernst Porsche, born in 1909, came to work for him, and together they pursued a passion for technology and innovation.

At this time, the German government, focused on economic and social development, was looking to partner with a company to create the "people's car," which would serve as a family car for the modern German household. With this inspiration, Porsche developed the Volksauto, later called the Volkswagen. The Volksauto combined a number of innovations of the day, including an air-cooled engine (in the rear), suspension, and a beetle-shaped hood, designed for better aerodynamics. In many circles, it was believed that Porsche had created the perfect car for the job at hand: transporting a family of two adults and three children at up to 60 miles per hour.

The 1940s brought global conflict in the form of World War II. Summoned to provide vehicles that were needed for the current situation, Porsche began to develop military vehicles. Porsche introduced the Kubelwagen and Schwimmwagen, which were both militarized versions of the Volkswagen Beetle. Porsche then designed a number of tanks for use in World War II, ultimately designing the chassis for the German Tiger Tank (Tiger I).

In 1945, when war ended, Ferdinand Porsche was arrested by French soldiers for his role in the conflict. The 22 months he would spend in prison allowed his son, Ferry Anton, to create a new vehicle designed for racing, named Cisitalia. This was a preview to the first Porsche sports car, which would be introduced in 1950. And again, we see the Porsche family with a keen sense of design for purpose, building the vehicles that the situation dictated.

THE PORSCHE SPORTS CAR

When he emerged from prison in 1947, Ferdinand Porsche marveled at the work that his son had done in his absence. Upon examining the design and prototype, he remarked, "I would have built it exactly the same, right down to the last screw." With Ferdinand Porsche beaming with pride and his company flush with preorders for the new innovation, the Porsche model 356 was produced and then road certified in 1948. It was a mere 40 horse-power, but sleek handling, comfort, and reliability (rare in sports cars of the time) distinguished the design.

The Porsche company, with a history of innovation, went on to introduce its own engines in the mid-1950s, leading up to the iconic Porsche 911, intro-duced in 1963. As the successor to the 356, the Porsche 911 built upon much of its original design, but with a six-cylinder boxer engine, an internal combustion design.

In the time since its first foray into racing, Porsche cars have won over 24,000 races globally, with the greatest notoriety coming with more than 50 class wins at Le Mans. A company that started by building a family car evolved into military vehicles, and then reinvented itself as perhaps the greatest sports car company ever.

PORSCHE TODAY

Unlike many of its competitors, Porsche still operates as a standalone entity, profitable and renowned for design and performance. While the company could have been content with its place as number one in sports cars, the innovative culture would not rest. It always saw another job to get done and sought to design a vehicle to complete that job.

The sports-car line has evolved from the Porsche 911 to include the Boxster (a two-seater), the Cayman (a two-seater, evolved from the Boxster), and many varieties within each line (six- and eight-cylinder, cabriolet, and so on). Perhaps the company's most aggressive bet came in 2002, with the introduction of the Porsche Cayenne. As Porsche's first sports utility vehicle, this model broke the mold on 40 years of innovation. The Cayenne, sporting five seats and an expanded trunk area, was designed to cater to the current needs of growing families. While the market response was initially muted, the Cayenne quickly became Porsche's largest seller in terms of volume.

But, the innovation did not stop there, with the introduction of a sedan (the Panamera) and most recently, the Macan, a smaller crossover sports utility vehicle. While each vehicle possesses supreme performance and handling, each is unique to a certain class of buyer. It's customization of a different ilk. A key part of the engineering strategy has been to leverage common parts across the product lines, to drive efficiencies. This strategy enabled the company to deliver to many different client needs, at a value on par with the quality.

ROME (CONTINUED)

It's hard to forget the indelible image of the storekeeper, who had spent over 50 years of his life designing and creating belts. His passion, attention to detail, and focus on fit for purpose are enduring to this day.

That storekeeper, like Ferdinand Porsche before him, believed that satisfaction came from designing something that was perfect for the job at hand. As Part III will disclose, this fit-for-purpose ideal is a fundamental component of a data strategy for the Data era.

PUMA

HERZOGENAURACH

IN 1924, SITUATED in a small town called Herzogenaurach in Germany, two brothers named Adolf and Rudolf Dassler founded Gebrüder Dassler Schuhfabrik (Dassler Brothers Shoe Factory). Adolf ("Adi") Dassler was the skilled craftsman who designed and manufactured the shoes, and the older brother, Rudolph ("Rudi"), was the extroverted salesman. The company's visibility skyrocketed when Jesse Owens won four gold medals wearing Dassler shoes in the 1936 Olympics. However, the brothers did not see eye to eye, and many disagreements started to surface. Tension between the brothers was such that they eventually decided to split up the business in 1948. Rudolf moved to the other side of the Aurach River to start his own firm, initially called "Ruda," from "Ru" in Rudolf and "Da" in Dassler, before eventually renaming it to Puma. Adolf also started his own company called Adidas using his nickname, Adi, and the first three letters of his last name. Thus the rivalry between Puma and Adidas was born.

This rivalry was not a simple case of business competitiveness, but rather a bitter battle between two brothers who seriously disliked each other. The small town of Herzogenaurach was split in half by the feud, earning the nickname of "the town of bent necks," as people constantly looked down to see which shoes were worn as a sign of allegiance to either Puma or Adidas. Local storytellers recall that handymen working at Rudolf's home would deliberately wear their old Adidas shoes in the hope that Rudolf would tell them to go to his basement to take a free pair of Pumas.

ADVERTISING WARS

The battleground for Puma and Adidas revolved around convincing star sports personalities to wear their shoes. Puma had great success with their campaigns. During the first soccer match after World War II, several members of the West German national soccer team wore Puma boots, including the scorer of West Germany's first post-war goal, Herbert Burdenski. Four years later, at the 1952 Summer Olympics, 1500-meter runner Josy Barthel of Luxembourg won Puma's first Olympic gold in Helsinki, Finland. Rivalry between the Dassler brothers came to the fore when German sprinter Armin Hary reached the 100-meter sprint final at the 1960 Summer Olympics and both Puma and Adidas competed to have Hary wear their shoes. Hary (hoping to cash in from both firms), having formerly worn Adidas, switched to Puma as he raced to win gold and then donned Adidas shoes on the podium when receiving his medal.

The rivalry between Puma and Adidas looked like it might be toned down with the next generation when Armin Dassler (Rudolf's son) and his cousin, Horst Dassler (Adi's son), signed an agreement known as The Pelé Pact, stating that neither Adidas nor Puma would enlist the famous soccer player. However, at the start of a 1970 World Cup finals match, Pelé stopped the referee with a last-second request to tie his shoelaces before kneeling down to give millions of television viewers a close-up of his Pumas. As a result, Horst was outraged and the pact between the companies fell apart.

In 1989, Rudolf's sons Armin and Gerd Dassler sold their 72-percent stake in Puma to Swiss business Cosa Liebermann. This paved the way for new leadership and recognition that the real competition in the market was coming from Nike and Reebok.

JOCHEN ZEITZ

Between 1993 and 2011, Jochen Zeitz was the CEO and Chairman of Puma. Zeitz was the youngest CEO in German history to head a public company at the age of 30. He masterminded the worldwide restructuring of Puma, which was suffering from major financial difficulties at the time. His long-term development plan was responsible for a 4,000-percent increase in Puma's share price over the course of 13 years, from $10.86 in his first year as CEO to an all-time high of $442 when the majority stake of the company was acquired by Kering (previously known as Pinault-Printemps-Redoute, or PPR) in 2007. It is remarkable that Zeitz managed to transform Puma from a low-priced, relatively undesirable brand into one of the top three brands in the sporting good industry.

Zeitz is a true business leader and is not afraid to go where others will not venture. He has an impressive ability to spot future trends and seize opportunities before the competition arrives. In 2008, he created PumaVision, an ethical framework defined by the four key principles of being Fair, Honest, Positive, and Creative, as applied to all professional behavior, business procedures, and relationships throughout and outside of Puma.

ENVIRONMENTAL PROFIT AND LOSS

Possibly the biggest innovation in the business world coming from Zeitz has been his challenge to other business leaders to recognize their use of ecosystem services across the supply chain. Rather than simply preach about the environment, Zeitz realized that Puma's customers had a growing desire to know that their purchases were not harming the environment. Therefore, Puma addressed the challenge of assessing the environmental impacts of a product at each stage of the product life cycle — from the generation of raw materials and production processes all the way to the consumer phase. Big data is at the heart of the process for measuring these environmental impacts.

Through the leadership of Zeitz and colleagues, Puma introduced an Environmental Profit and Loss (EP&L) accounting system in 2011 that measures the true costs of the business's impacts on nature, by calculating the monetary value of the company's greenhouse gas emissions and water consumption across its entire supply chain. The EP&L serves to help firms combine sustainability metrics with traditional business management. Rather than fall prey to green washing tactics and negligible action, the EP&L allows managers to develop strategies for sustainability that can actually be operationalized. In addition, the EP&L accounting system provides an overarching metric to assess risk and opportunity across operations, products and supply chains. According to Puma, the EP&L has been indispensible to realize the immense value of nature's services that are currently being taken for granted but without which companies could not sustain themselves. The EP&L helps Puma is to reduce its impact on the environment (the "loss" in an EP&L) as far as possible while continuing to deliver value to its customers. Puma claims that calculating these impacts will help it reduce costs and develop a more sustainable and resilient business model by safeguarding the resources and ecosystems upon which it depends for long-term success.

Clothing and sporting goods manufacturers such as Puma are among the most prominent exponents of the outsourced production model, and hence the EP&L is important for obtaining a true picture of the cost of producing its goods in terms of the natural resources consumed. The environmental impact for the key areas of greenhouse gas (GHG) emissions, water use, land use, air pollution, and waste generated through the operations and supply chain of Puma was valued at $183 million in 2010. Only $10 million of the $183-million total derive from Puma's core operations such as offices, warehouses, stores, and logistics, while the remaining $173 million relate to Puma's supply chain. The biggest contributors to the total impact of $183 million were GHG emissions and water consumption with an economic valuation of $119 million.

What is rarely apparent is that the success of Puma's EP&L relies heavily on big data about a multitude of processes, such as water abstraction, energy use, greenhouse gas emissions, air pollution, waste disposal, and land use. Trucost pioneered the valuation of environmental impacts and provided the capability to model complex interactions across the supply chain to facilitate the EP&L accounting system. Without appreciating the importance and relevance of the data that Puma had collected about how the sources of value were distributed across its supply chain, it would never have been possible to conceive of the EP&L.

Interestingly, the success of the EP&L is likely to hinge on access to more data and the development of big data analytics. A panel of experts were asked by Puma's parent company, Kering, to review the EP&L and identify where potential improvements could be achieved. The panel suggested some technical areas around valuation where the methodology could be improved. In particular, the panel acknowledged the importance of big data and recommended increasing the amount of primary data on which the EP&L is based in order to reduce the levels of uncertainty. At present, the calculations required for the EP&L rely significantly on estimation techniques such as environmentally extended input-output (EIO) modeling and sourcing location information is limited to the country level.

Zeitz is continuing with his goal of encouraging business leaders to move towards proper accounting for environmental impact by adopting Puma's EP&L accounting system through his activities within the B-Team, which he co-founded and co-chairs with Sir Richard Branson. The B Team is a not-for-profit initiative that brings together influential global leaders and international business CEOs to create a future where the purpose of business is to be a driving force for social, environmental, and economic benefit.

HERZOGENAURACH (CONTINUED)

The rivalry between Puma and Adidas continued for several decades and reflected the animosity that existed between the two Dassler brothers. It was only in 2009, after both brothers had died, that the two companies made a truce and organized a friendly soccer match. Sadly, the two brothers never made peace, and although they are buried in the same cemetery, they are as far apart from each other as possible.

A METHODOLOGY FOR APPLYING BIG DATA PATTERNS

INTRODUCTION

THE DATA ERA introduces a significant cultural change to every organization and individual: It's a shift from intuition-based decision-making to fact-based. While this seems simple, it is much harder in practice, as it causes a company or individual to unlearn their traditional approach. Intuition was sufficient and necessary, when data was limited. But, in a data-rich environment, there is no excuse for anything but thoughtful decision making, grounded in data.

Chapter 2 uncovered how this change in approach is transforming medicine. With the advent of the Data era, physicians are forced to rethink their opinion-based approach and adopt a data-based approach. This will likely lead to one of two outcomes:

1. Doctors will make the adjustment, fully incorporating new sources of data in their decision-making. This will improve the quality and cost of medical care, while requiring physicians to expand their skills beyond just the practice of medicine.

2. Doctors will become irrelevant, in many cases, replaced by intelligent machines and monitoring which will make treatment decisions based on the flow of data and enable self-management of many medical disorders.

The shift to data-based decision-making is a cultural change. A methodology rooted in a fact-based approach becomes a helpful tool for driving that cultural change.

THE METHOD

The method for applying big data patterns to decision-making consists of seven steps. While applying the method does not require every step every time, it is suggested that each step be considered for its potential impact. Because this is a cultural change, the steps required may not always be obvious.

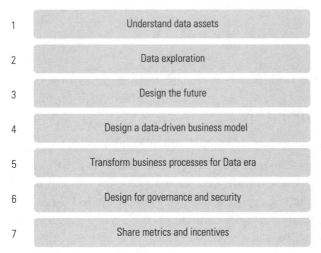

1	Understand data assets
2	Data exploration
3	Design the future
4	Design a data-driven business model
5	Transform business processes for Data era
6	Design for governance and security
7	Share metrics and incentives

A methodology for applying big data patterns

The methodology is intended to be practiced sequentially. Each step builds on the one before, demanding a fact-based approach at each interval along the way. The effort applied in any given step depends a bit on the end goal. In some cases, the goal is simply analyzing existing data assets, which would drive the majority of the focus to Steps 1 and 2. In other cases, a company may be starting a new business based on new data assets, which will demand detailed attention to all steps. And in other cases, a company may define the core need around redesigning a key business process based on data assets, which would put the emphasis on Step 5.

Here is an overview of the key steps and a summary of the intent for each component of the methodology:

1. **Understand data assets.** Locate, catalog, and assess the potential of, all relevant data assets. These assets could be inside or outside the organization.

2. **Explore the data.** Explore the data assets discovered in Step 1, applying a rigorous methodology of searching for statistical significance.

3. **Design the future.** Use the implications of the data exploration to either make better decisions, design a new business model, or redesign current business processes.

4. **Design a data-driven business model.** If appropriate, design a business model to take advantage of the new data discovery.

5. **Transform business processes for the Data era.** If appropriate, redesign existing businesses processes to capitalize on the new insights from Steps 1 and 2.

6. **Design for governance and security.** Understand the impact of leveraging new insights and data assets, comprehending the implications of privacy and data usage.

7. **Share metrics and incentives.** Develop a system to ensure that appropriate key performance indicators are measured and that stakeholders continue to leverage fact-based decision-making or utilize new findings.

STEP 1: UNDERSTAND DATA ASSETS

Many traditional organizations view data as being peripheral to the core business strategy. Data simply serves as a record of historical business transactions, and regulators may advise the firm to keep these records for a number of years in case a need should arise to audit or investigate a dispute. In such organizations, storing and managing this data is often treated as an overhead cost, and it is rarely seen as a source of information that could influence the future of the firm. In fact, it is often the case that managers remain completely disconnected from the information technology (IT) department that is responsible for the data. This disconnect is typically exacerbated by a lack of communication between the technical experts who are tasked with the job of making sure IT systems function from day to day and those in management who are responsible for long-term strategy.

Thinking of data as an asset calls for a shift in thinking. It may be hard for managers without a data science background to fully grasp the potential value of the data collected within their own organization.

Many organizations are realizing the value that certain types of data can bring, and even if it cannot be monetized internally, they are identifying partners who are confident of converting the data into financial rewards.

THE PATTERNS

Twelve of the 54 big data patterns previously revealed in this book capture the key aspects of the first step, understanding data assets.

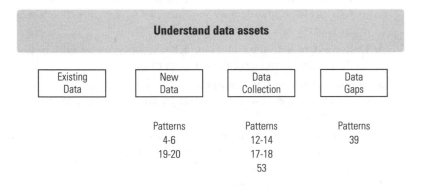

Existing Data

While one would hope that within each organization there is at least one individual who is aware of the different types of data that already exist, in reality, this is rarely the case. It is usual to find that, due to organic evolution of IT systems and successive upgrades, a series of different databases will coexist and be used for different functions. For this reason, even the task of being able to cross-reference between all the existing databases may be a substantial challenge.

New Data

Faced with the task of including new datasets in an organization, it is important to think carefully about the challenge of integrating existing and new data. Again, the focus should be to insist on a facility for allowing queries and applications to run over all the databases without intervention from human experts. Once data is viewed as a valuable commodity with potential for growing the business, as opposed to a painful task that has to be implemented, there will be more willingness to design data architectures that are appropriate for data science.

The identification of relevant new data will often require unique thinking and may involve searching amongst non-traditional sources. Being aware of what competitors are doing is one way to avoid being left behind, but it's certainly not a means to achieve a reputation for being innovative. It is generally wise to reach out to data experts who are familiar with the challenges that the organization is trying to address when seeking new sources of data.

There are many characteristics of data that should be considered when evaluating their fit for analysis. These include:

- Sampling frequency
- Spatial coverage
- Accuracy
- Reliability
- Timeliness
- Price

Undertaking a preliminary study to assess the advantages and disadvantages of different datasets with respect to each characteristic is recommended before making a final decision.

Patterns 4 through 6, and 19 and 20 demonstrate the impact of new data:

4. Creating new sources of data may lead to competitive advantage. Most companies have a current source of data and do not think about external sources where they can obtain new data.

5. Joining previously disconnected or disparate datasets can lead to new insights and competitive advantages.

6. New data types emerge from the fusion of internal and new external sources of data.

19. Data from social networks provides a diverse set of opinions, enabling more exhaustive analysis and often better outcomes.

20. Data from social networks serves the purpose of challenging opinions from the experts in the process.

Data Collection

Obviously, the ingestion of data into an organization should be driven by a data science strategy. Before purchasing new databases, a cost benefit analysis should be undertaken and attempt to forecast as far into the future as is feasible. Many external sources of data will already exist, and it is usually possible to obtain samples before committing to paying for the service.

It is important to think carefully about the value of having live, as opposed to historical, data. If decisions are to be made in near real-time and the cost of delay is substantial, then the inclusion of data will most likely involve an Application Programming Interface (API). APIs are commonly used to access social media platforms such as Facebook and Twitter.

Patterns 12 through 14, 17, 18, and 53 indicate the importance of data collection:

12. Collecting and synthesizing data will be a core competence of leaders in the Data era.

13. Data that is collected must be transmitted and analyzed in a timely fashion.

14. Data must be collected over the lifecycle of a company, product, or business process. Data at a point in time is often interesting, but not necessarily complete.

17. Some data is time sensitive. Time-sensitive data must be collected in a continuous manner, producing a data stream.

18. Time-sensitive data must be analyzed using data-stream analytics as that data is generated so that it can be acted upon.

53. Novel sources of data can help to monitor and respond to risks.

Data Gaps

It would be wonderful if nobody ever made a mistake when taking measurements or if sensors never malfunctioned. In practice, all datasets contain gaps, and a data scientist's first task is to decide how to deal with such gaps.

In fact, while gaps are a nuisance, at least it is clear that these values have to be *imputed,* the term given to filling in missing values. A bigger challenge occurs when one is suspicious about certain data values that seem to be much bigger or smaller than expected. These values are known as *outliers,* and a systematic approach is needed for detecting them. In some cases — for example, negative sales — it is easy to identify these outliers as measurement errors. However, in many cases, some human judgment may be required to make a decision about whether to accept or reject these potential outliers. For example, extreme values could be either a measurement error or simply a new record high or low. Knowledge about the system that generates the data is helpful for knowing what to expect and hence to decide whether or not to treat such values as outliers. In situations in which no scientific model is available, it may be necessary to use a statistical model based on historical data to label outliers.

The task of imputing missing values is as complex as constructing a model for the entire dataset. In one respect, filling data gaps can be viewed as creating data, and hence the validity of such data is questionable. If at all possible, it is certainly best to avoid creating data and then using this data for making decisions. However, for many quantitative models, it is necessary to have a continuous time series, and for this reason, selecting an approach to filling gaps cannot be avoided.

The simplest approach is to fill a gap with the last reliable observation. This is a common approach for time series that do not change much between one observation time and the next. Furthermore, it has the advantage of not introducing any bias that could be used to provide an unrealistic level of forecast accuracy. For time series that vary a lot between observation times, it may be desirable to use linear interpolation to fill gaps. This may also be necessary when gaps cover multiple observation times. But beware: The use of linear interpolation introduces a linear bias in the data because information from the past has been propagated into the future by the gap-filling approach.

It is recommended to use only actual observations for calculating performance metrics that are based on averages calculated across datasets. In this way, while gaps have to be filled to enable the quantitative model to be estimated and produce forecasts, the new data that has been created is not permitted to affect the final results.

Pattern 39 speaks to the importance of data gaps on innovation:

39. Determine where the lack of data is extending the innovation lifecycle and eliminate the constraint.

STEP 2: EXPLORE DATA

The strategies deployed by many successful organizations are often based on intuition, knowledge, and an expectation that what worked previously will continue to work. Big data brings an ability to turn decision-making into a scientific exercise that is easy to repeat, justify, and explain to independent agents.

It must be acknowledged that data does not and cannot be expected to capture every nuance of the workings of a complex organization. This does not excuse the over-reliance on gut feelings for decision-making simply because it is believed that the use of data is pointless. In contrast, there are many situations where data and quantitative modeling can be used as a means to communicate complicated ideas. When used properly, data-based modeling can indicate when the uncertainty is overwhelmingly large so as to negate the potential for providing any concrete results. This honesty about the levels of ignorance is rarely a characteristic found in business leaders. On the whole, the discipline required for modeling helps to make decision-making more transparent and minimizes the risks of relying on assumptions that cannot be supported.

A good starting point for those wishing to experiment with data science is to focus on some challenges that company management is grappling with and attempt to create a list of working and future assumptions *(hypotheses)* about these key aspects of the business. The following figure illustrates what such an exercise might look like. The linear flow chart provides the big picture of going from data to decisions. In contrast, the circular flow chart describes the more routine task of using data science to address business challenges. The components of this data science approach are discussed in the following section.

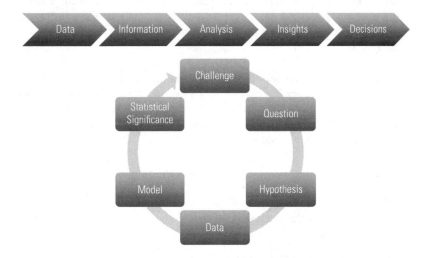

CHALLENGES

The identification of a roadmap for testing the efficacy of a data science approach to informing decision-making within an organization requires some careful consideration of what aspects of its operations stand to gain most from imposing the discipline that comes with quantitative analysis. By considering some challenges already being experienced by the organization, it should be possible through some soul searching and questioning to convert these challenges into testable hypotheses.

QUESTIONS

One of the main reasons for having a board of directors is to obtain a diversity of opinions about what is best for the organization. When attempting to forecast the future, such diversity is even more important. In many situations, trends may already be taking hold and empirical data could be used to test for the existence of trends and to quantify the strength and duration of such trends. Furthermore, it should be possible to ask questions about the impacts of emerging trends and how these will influence key performance indicators chosen by the organization.

HYPOTHESES

Indeed it would be healthy if directors who hold conflicting views could be sufficiently open to an empirical experiment based on data as a means of rejecting or accepting some of these views. Being able to communicate opinions and ideas is an important attribute of any manager. It will become increasingly desirable to offer such views in terms of hypotheses and to suggest ways for testing these empirically.

DATA

The empirical aspect of this approach relies on the ability to identify and access relevant datasets that could be used to test these hypotheses. This is where the bottom-up (data-to-management) and top-down (management-to-data) approaches should be replaced by a circular pathway of connectivity. Unless both management and the IT staff in charge of collecting and storing data recognize the importance of data science, it will be impossible to be successful. Both internal and external data should be considered as a means of testing the hypotheses of the management team.

MODELS

Quantitative models should be selected and constructed to test the hypotheses. An appropriate model will depend not only on the available data, but also on the hypothesis that is being tested. A wide range of models exist that could be deployed. For example, is a linear description adequate, or should one consider nonlinear models that could be identified using machine learning approaches? Where there are known interactions between variables, it is advisable to consider machine learning techniques. Another consideration is the use of parametric or non-parametric models. Parametric models are generally described by mathematical equations that require estimation of a finite number of parameters. They are relatively easy to explain and share within the organization, as only the equations and parameters need to be communicated. Non-parametric models usually rely on access to the full database, and for this reason, they are less convenient for transporting.

The approach used for estimating and evaluating a model should be designed such that it is possible to reach an outcome whereby the data is deemed insufficient to reliably answer the question that was posed. This can be achieved by seeking consistency between the model and the data, and by quantifying the level of confidence in each estimate. In situations of high uncertainty conveyed by low confidence levels, such outcomes would suggest that the quantitative approach is not able to deliver any insights.

STATISTICAL SIGNIFICANCE

Finally, the empirical evidence is presented in a statistically rigorous format. This approach will move the debate about major decisions into the Data era, where not having empirical support for an argument will be viewed as a weakness. Intuition is still an important aspect of hypothesis generation, but it is strongly recommended to collect appropriate datasets and to attempt to test ideas empirically.

Demonstrating an ability to reject certain hypotheses and communicating the levels of statistical significance will offer varying levels of support for various assumptions that can be clearly conveyed to the management board. It is also apparent that a multidisciplinary team will be required to facilitate the quantitative and qualitative aspects of this approach to data-driven decision-making.

STEP 3: DESIGN THE FUTURE

Even in the fact-based Data era, imagination is required. Designing the future demands creativity and intuition, based on data, to envision the opportunity to capitalize on new data assets or insights.

When comprehending the future, a business must think about all stakeholders in the business. It's not just clients, although clients are critical. It's not just suppliers, but they could be essential. It's not just employees, but employees are a critical stakeholder.

There are three constituents that should drive thinking and behavior when designing the future:

- Clients
- Employees
- Users

They are all equally important. At times, they could be similar; and at other times, they could be unique. A business must understand how to make it easy for each constituent to work with the products and services produced by the company. With each new idea, a leader must be able to articulate the must-have experience and the target of that experience (clients, users, or employees). Designing the future requires a rigorous process for identifying the most passionate stakeholders and getting their unstructured feedback.

THE PATTERNS

Ten of the big data patterns previously revealed in this book capture the key aspects of Step 3: Design the Future.

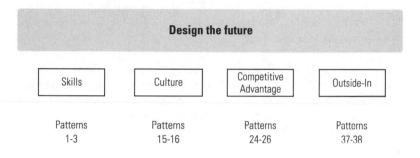

Skills

Patterns 1 through 3 demonstrate the skills demanded by the Data era:

1. Understanding the skill sets needed today, tomorrow, and in the future, based on the potential for data disruption.
2. Redefining roles and skill sets to take advantage of the new data available that can impact business processes.
3. Training and retraining current and new workers so that your business can remain relevant.

An organization must accept that the skills they have today may not be sufficient for the Data era. The challenge is that the skill gaps, once understood, will likely prove to be excessively broad. The Data era demands skills in data science, statistics, and probability. And that's just on the business side. On the IT side, the skill needs are much different than traditional IT, with a premium placed on programming and modeling skills.

Leading organizations will document their skills today versus the skills needed tomorrow and systematically begin to fill the inevitable gaps.

Culture

Patterns 15 and 16 demonstrate the cultural challenges posed by the Data era:

15. An individual or organization must document their biases (or sacred cows) up front to ensure those biases are recognized and understood.

16. Decisions will start with data, not opinion or organizational sacred cows. A simple test to ensure this is happening is to force every decision to be supported by data and carefully documented for future analysis. If it cannot be, then you can explore why (i.e. was the data uncollectable, unavailable, un-analyzable?).

Even those organizations that begin with Steps 1 and 2 in the methodology will face cultural resistance to moving to the Data era. It's human nature. The only way to account for these biases is to document them up front. It will be important for the organization to be aware of existing cultural biases, as awareness becomes the road to mitigation.

Competitive Advantage

Patterns 24 through 26 demonstrate the opportunity to create competitive advantage in the Data era:

24. Competitive advantage can be gained by leveraging data or by monetizing data assets.

25. Data must be consumable by external parties if there is a goal of monetization.

26. Companies should not monetize data that erodes their existing competitive advantage.

The competitive advantage begins with imagination and creativity playing a prominent role. Experienced insight, based on data, can help a business decide when to take opportunities to leverage or monetize data, and when not to. Effectively understanding and building a strategy based on a competitive advantage will require cross-functional participation because an advantage in one place could create a disadvantage in another.

Outside-In

Patterns 37 and 38 demonstrate the benefits of outside-in thinking and how that can lead to new partnerships or opportunities in the Data era:

37. New partnerships will emerge as a result of innovation driven primarily by data.

38. New insights arise from combining the assets of separate companies, starting with their data assets.

As previously mentioned, the Data era will usher in new partnerships that would have never been conceived in prior eras. This starts by understanding the ecosystem of companies that serve the same customer base and clearly delineating how each company provides their value proposition. With this knowledge, a company can determine if a third party is really complementary or potentially competitive. In the case of IBM and Apple, as mentioned in Chapter 12, Apple viewed their competitive advantage in devices. IBM, having no device business, saw how devices, with the right data and applications, would be more valuable to enterprise clients.

STEP 4: DESIGN A DATA-DRIVEN BUSINESS MODEL

Data can be the basis for building a new business, division, or partnership. In fact, as we've seen with companies like CoStar and IMS Health, data can form the crux of the business model. However, this approach may not be for every company. Some companies, like Inuit, may decide that data will enable competitive advantage for their products and services. Understanding which direction to take is the essence of Step 4: Design a Data-Driven Business Model.

THE PATTERNS

There are seven of the big data patterns previously revealed in this book that capture the key aspects of a Design A Data-Driven Business Model.

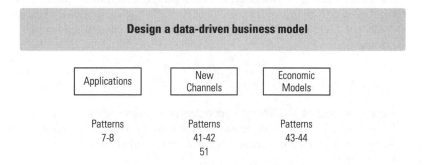

Applications

Patterns 7 and 8 illustrate the leverage that a business can create by applying data to new or existing applications and business processes:

   7. New data applications will enable faster and more productive use of data.

   8. Data applications will evolve as data sets evolve to ensure continuous insight and learning.

Perhaps the best illustration of this is the emergence of smartphone applications and how they are impacting both business and personal lives.

Smartphone applications are primarily data applications. While they provide other functions, the applications with widest appeal (outside of games) are those that deliver data in a consumable form. Examples that come to mind are maps, Twitter, and Facebook. Every organization has the data to provide an engaging relationship with customers and suppliers, if they choose to.

New Channels

Patterns 41, 42, and 51 demonstrate the power of developing new channels and partnerships on the foundation of data:

   41. New channels to market can change the economics of an industry.

   42. Data can unlock new channels to customers.

   51. Cooperation between the public and private sector to create innovation from big data.

Data applications can foster new relationships, as demonstrated in Patterns 7 and 8. These new relationships can create access to new channels and markets, accelerating growth for a business. A fresh perspective, from an outsider, is often needed to explore the opportunity in new channels. Otherwise, a business can be encumbered by the historical definition of their industry, market, and channels.

Economic Models

Patterns 43 and 44 highlight the opportunity to realize new economic benefits based on data. This benefit can come in the form of new products or services, enhanced products or services, or even entirely new business opportunities:

43. Data will create new economic models, often providing the ability to better align customer and supplier interests.

44. New pricing or economic models, based on data, can disrupt incumbent and entrenched suppliers.

STEP 5: TRANSFORM BUSINESS PROCESSES FOR THE DATA ERA

Data provides the opportunity to eliminate, transform, or define new business processes to capture value.

THE PATTERNS

There are 18 of the big data patterns previously revealed in this book that capture the key aspects of Transform Business Processes for the Data Era.

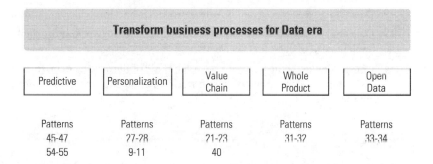

Predictive

Data has predictive powers. Although data analysis has traditionally been a retrospective view of what happened in the past, we now have the capability to predict and model the impact of future events. This is now possible due to the ability to collect more data, along with the refined tools and techniques for performing sophisticated analysis. With this new predictive power, data assets can be utilized to eliminate, transform, or create new business processes:

41. Data provides the basis for predicting future events, outcomes, and impacts.

42. Forecasting and predicting is not valuable unless it is acted upon. Data often holds the answer to what action should be taken.

43. Predictive analytics require the fusion of historical and real-time data, along with internal and external data.

54. Construct early warning systems for the most dangerous risks.

55. Being prepared and having action plans in place allow rapid responses to early warning systems.

Personalization

Patterns 27, 28, and 9 through 11 demonstrate how data has transformed customer engagement from a segment-based approach to a personalized approach. A company can now treat each customer as they are: an individual, with unique tastes and a desired approach for engaging with the company. This causes a company to build business processes that assume personalization, which is something that their original processes could never help them comprehend:

27. Data will drive the personalization of sales and marketing.

28. With personalization, data can change the economics of many industries and business processes.

9. Data will transform current business processes and introduce new ones.

10. Data enables the personalization of business processes to an individual level, instead of customer segments.

11. Companies are being born solely on the basis of harnessing data, to disrupt incumbent companies with traditional business processes and approaches.

Value Chain

Patterns 21 through 23 and 40 capture how the value chain of a business will be transformed by data. This will eliminate entire steps and constituents in the name of better efficiency and service. Data drives specialization, which means that information aggregators are no longer required. This causes a disruption in many value chains:

21. Utilize data to remove steps in the value chain and improve the end product.

22. Understand how data can transform, expedite, or reduce the cost of steps in the value chain.

23. Specialists are frequently paid more, driving a free-market desire for more specialists and in turn, more biases.

40. Re-examine the innovation lifecycle to automate or remove parts of the cycle with the application of data.

Whole Product

Data facilitates the creation of long-term relationships between parties. Patterns 31 and 32 capture this phenomena. The reason is simple: They come to know each other better, driving a greater codependence. Therefore, the whole becomes greater than the sum of the parts. Data is a unifying asset in this respect:

31. With data, the definition of a product may extend beyond features and function to experience and serviceability.

32. Data can change products from being transactional in nature to iconic and desirable, offering long-term relationships with a customer.

Open Data

Data is being democratized. Like many revolutionary movements, it starts at the bottom. Customers, partners, and employees are demanding greater transparency and information sharing. In this era, a primary role of an enterprise is to democratize access to data, sharing it appropriately to foster the right relationships. Patterns 33 and 34 highlight the benefits of open data:

33. The democratization of data is starting to happen, putting an impetus on companies to organize and govern the data made available.

34. Data search answers only questions that the user knows to ask. So, effective access to data must serve up the right data to the right person at the right time.

STEP 6: DESIGN FOR GOVERNANCE AND SECURITY

Data in the right hands is a productive asset. As previously described in this book, it can foster new relationships, encourage long-term loyalty, and transform business processes. In the wrong hands, data is useless or perhaps even dangerous. Bad data leads to bad outcomes, whether its in the right or wrong hands. All of this puts the impetus on data governance and security. With the recognition of the value inherent in an organization's data, it's apparent that this can no longer be an afterthought.

THE PATTERNS

There are four of the big data patterns previously revealed in this book that demonstrate Step 7: Design for governance and security.

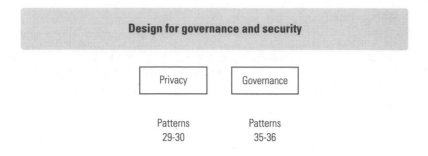

Privacy

Patterns 29 and 30 illustrate the catch-22 of a data-rich world. More data enables personalization and better business relationships. It also has the risk of sacrificing privacy. Therefore, data policies and procedures become critical to ensure that data trust is not violated:

29. Privacy and personalization are often at odds with each other.
30. Understanding how to utilize personalization, without overtly sacrificing privacy, is critical to appropriate data stewardship.

Governance

Data governance captures who and what accesses data, how that data is used, where it's moved to, and how it is ultimately archived or disposed. This entire lifecycle of managing data is a requirement in the Data era. Without this capability, a business will likely not remain a going concern, as data leakage is inevitable without these protections. The challenge with data governance in an enterprise is that the greatest threats are not external; they are internal, within the organization:

35. Security of data is just as important as leveraging data for insights and business transformation.
36. Data governance is necessary to manage the lifecycle of data assets so that you can ensure data security and ensure compliance with business or legal practices.

STEP 7: SHARE METRICS AND INCENTIVES

There is an old saying, "You get what you measure, not what you expect." It's a classic view on management and leadership, highlighting the fact that without knowing where you are going as an organization, you are likely to never arrive.

There are three of the big data patterns previously revealed in this book that highlight the importance of metrics and incentives:

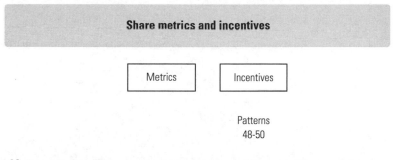

48. Incentives will frequently determine the willingness of an organization or individual to take risk.

49. A proper incentive framework may determine success (or failure) in an initiative to apply data to a business problem.

50. Organizational boundaries or silos lead to sub-optimal outcomes and insights. Incentives must break down this barrier.

While it's easy to understand in concept, defining revlevant metrics and aligning incentives is often easier said than done. But when this is done right, organizational boundaries disappear, and all stakeholders start to move in the same direction.

BIG DATA
ARCHITECTURE

INTRODUCTION

THE METHODOLOGY PRESCRIBED in Chapter 16 is a framework for applying the patterns in big data revealed throughout Parts 1 and 2 of this book. A methodology is much closer to a recipe for cooking than it is a recipe for baking. When you are baking, you must adhere precisely to the recipe. It is a precise prescription. Even the slightest deviation can cause the outcome to vary dramatically. In contrast, when cooking, a recipe is guidance, which can be adjusted based on your sense of smell, taste, and appearance. Cooking is an art, while baking is much more of a science. The methodology in Chapter 16 is closer to art, although it must be thoughtfully applied.

ARCHITECT FOR THE FUTURE

The architecture portion of the methodology is part art and part science. Common architectural patterns and taxonomy have emerged from a variety of customer experiences. This commonality facilitates understanding the proper end state, as well as the logical progression path. Getting from point A to point B is often more important than simply arriving at point B.

Architecting for the future necessitates a uniform, repeatable, and proven approach to achieve the anticipated business value that big data promises. Business value is unlocked through the implementation of the patterns, and a proper architecture ensures that implementing the patterns is possible. The big data architectures introduced here and covered in depth in Chapters 18 through 20 are a series of architectural patterns that describe the best approach for enabling big data use cases. These architectures are based on proven customer success, as many have embarked on their journey to the data era.

That being said, most big data initiatives are implemented within the context of an existing data and information environment. This requires flexibility in approach, given that the starting point and even the steps along the way may be slightly different from organization to organization. This is why implementation is more analogous to cooking than it is to baking. It's just not as simple as precisely following a recipe.

The patterns in big data are modernizing existing traditional enterprise design patterns, with big data concepts, to enhance competitive value. Therefore, it's appropriate to plan for this to be an evolutionary approach, as opposed to a revolutionary one. The patterns and the supporting architecture articulate a to-be state, with each organization evolving at different rates of change and maturity.

Understanding the architecture of the future, as well as the priority patterns for each individual organization, will create a number of benefits for the adopting organization. The anticipated benefits include:

- **Reduced time to value:** Reduced deployment time and efficiency of execution and adoption
- **Reduced cost:** Reduced cost of deployment, based on more efficient use of resources
- **Reduced risk:** Given faster time to value, lower cost of getting started, and the reliance on a trusted methodology, risk is dramatically reduced. Leveraging the patterns and architectures described here directly benefits from the experiences of others. Those that helped define the 54 patterns (presented in Chapters 12 and 16) have paved the way.

Architecture is the step that realizes the potential of the patterns in big data. This is where the real work happens.

LESSONS FROM STUTTGART

In Chapter 14, Porsche taught the lesson of fit-for-purpose. Porsche did not have to expand their line of automobiles to offer anything more than the sports car that they eventually became known for. However, the lesson from Porsche, as well as the Italian belt maker, is that as different needs arise, different solutions are required. Understanding the job to be done is critical to understanding which approach to utilize to address that job.

A simple graphic illustrates the reference architecture for Porsche, based on the different needs in the market.

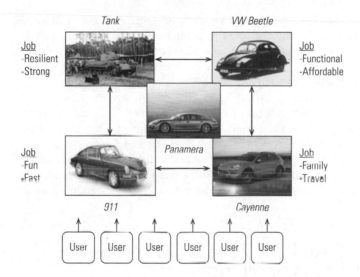

Each vehicle represents a different approach or capability to addressing the job that must be accomplished. In the case of VW Beetle, the job to be accomplished was very different than with the Porsche Cayenne. Ultimately, reference architectures are a means of distilling complex topics into something that can be easily understood and applied.

BIG DATA REFERENCE ARCHITECTURES

There are different styles of reference architectures, based on the topic and the audience. Some architectures focus on the business user, while others are very technical in nature. Each has its merits, based in a large part on the phase of the project and the role of the person utilizing the architecture. There are four key views to be achieved with a reference architecture:

- **Business View:** A detailed view of the capabilities required to achieve certain business outcomes. These capabilities are often presented from the perspective of a user, delineating how each capability contributes to executing a use case for a specific role or industry.

- **Logical View:** A view of the reference architecture required to deliver on the capabilities defined in the business view. Most people associate reference architecture with the Logical View, as described here.

- **Process and Data View:** An architecture depicting all business processes, the supporting data supply chain, and the actors in each part of the process. This would typically involve detailed process flows describing all steps in any given business process. This is very helpful in business-process reengineering exercises.

- **Physical View:** The Physical View of a reference architecture describes all systems, sub-systems and the requirements in each system. Requirements documented would be at the level of CPU, memory, storage, etc. for each system. The Physical View is necessary to achieve deployment success and is typically required before moving to deployment.

Chapters 18 through 20 focus on the Business View and Logical View of a big data architecture. These views enable an assessment of the capabilities needed and the framework of solutions necessary to deliver on those capabilities. As the Porsche example teaches us, those two views are often sufficient for understanding the landscape and planning for the future. These two reference architecture views will form a recommended architecture for the future.

LEVERAGING INVESTMENTS IN ARCHITECTURE

Before previewing the Business View and Logical View reference architectures, it's important to provide some context around short-term versus long-term planning in architectures. An enterprise must approach big data architecture with the long-term in mind. There are a number of approaches that create short-term gratification, but they will not scale to support an enterprise economically in the long term.

A simplistic view of enterprise reference architecture incorporates a data layer, application layer, and user interface. Simplistically, this can be represented graphically as a stack diagram (as shown in the following figure).

User
Interface

Application

Data Store

An enterprise IT organization invests across all three layers, as well as in integration between the layers. However, many emerging big data solutions are collapsing those three layers into one.

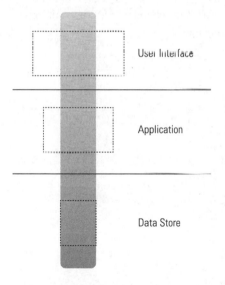

The gray bar represents a third-party product that provides the entire stack in proprietary packaging. This approach drives instant gratification. There is a single purchase; the enterprise gets the entire stack. New companies have been successful with this model. That being said, this is not a great approach for an enterprise in the medium to long term. Unfortunately, enterprises are not as simple as depicted in the preceding figure. They often look more like the following figure.

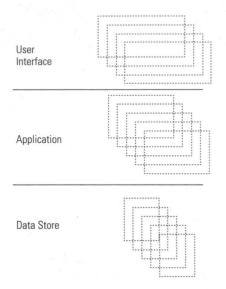

User
Interface

Application

Data Store

Enterprises are complex. They evolve, they buy, they build, and they customize. By taking the proprietary-stack approach to solving problems, an enterprise will eventually look like the following figure.

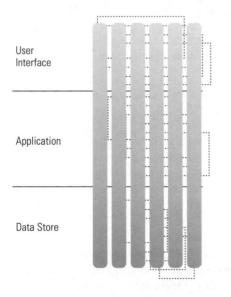

User
Interface

Application

Data Store

This is simply not economical for the medium to long term. With this approach, the enterprise ends up with tens and hundreds of full proprietary stacks (the gray bars) and no opportunity to innovate or customize. And, worst of all: The enterprise has no leverage in their investments. Each time the company requires a new capability, they have to re-buy capabilities they already have. They will start to regret their short-term gratification decisions. While some companies have been successful in the short-term with the instant-gratification model, it is not the right model for enterprises in the Data era.

Given the observations embodied in the 54 patterns of big data, an enterprise must have a flexible platform for managing data assets. Such a platform will leverage all information sources, annotate and organize that data, provide real-time analysis, and allow the enterprise to write their own applications or buy third-party applications. It will become the information bus for the enterprise. Like a school bus connects a number of unique stops and locations on its route, the information bus serves as the seamless integration point for access to all data assets. The following figure illustrates the leverage that this approach provides an enterprise.

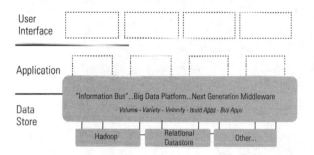

Intelligent enterprises do not make short-term gratification decisions en masse. They might make them to solve point problems from time to time. However, it is rare and doesn't reflect their strategy. Instead, the leaders in the Data era will take time to chart an architecture for the future, incorporating both a business view and a logical view.

BIG DATA REFERENCE ARCHITECTURES

The majority of companies will choose a strategy to leverage their architecture investments, as described earlier. With that as a backdrop, moving to implementation requires detailing out two reference architecture views: a Business View and Logical View.

BUSINESS VIEW

The Business View reference architecture delineates the component capabilities required for success in the Data era. This view does not consider sequencing or phasing. Yet it is possible to phase in these capabilities over time. It's best to think of the Business View architecture as a representation of all capabilities that may be required at some point. The Business View provides a bigger picture view, while the Logical View is a more detailed lens. Where an organization starts and how the organization progresses will be unique in most cases.

The business view reference architecture incorporates six major component areas, as represented in the following figure.

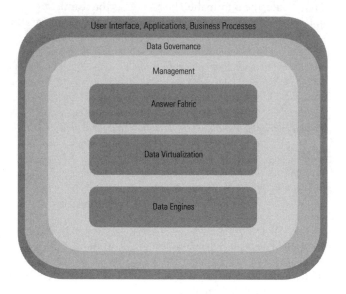

Each component of the Business View reference architecture must be comprehended as organizations evolve to the Data era:

- **User interface, applications, and business processes:** This is what the business users see as they interact with the big data infrastructure. User interfaces, applications, and business processes are tailored to the user, encouraging his engagement.

- **Data governance:** Ensures that data assets are managed like any other asset. Data assets must be secure, managed over their lifetime, and available for interrogation as required.

- **Management:** The management layer is a primary point of operation and orchestration for the IT team. This is where the architecture is administered and managed for optimal efficiency.

- **Answer fabric:** This is the brains of the data architecture. Analytical capabilities are leveraged to wow the users and drive business value to the organization.

- **Data virtualization:** Understanding that most organizations have to evolve from legacy architectures, data virtualization becomes a requirement to ensure all data assets are utilized. This capability prevents data from having to be moved to another location in order to be utilized.

- **Data engines:** If the answer fabric is the brains, the data engines are the nervous system, delivering the inputs to the brains so that the brain can take action.

Chapter 18 provides detail of each component in the Business View architecture, along with the key considerations in planning for each area.

LOGICAL VIEW

The Logical View reference architecture provides the technical detail to support the Business View reference architecture. This view provides a description of the technical capabilities required and how they will be materialized in the data architecture. The Logical View is a technical view, which can be directly translated into a Physical View at the right time.

The Logical View reference architecture incorporates five major component areas, as represented in the following figure.

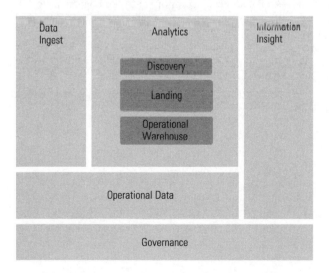

The components of the Logical View ensure that an organization is capable of executing on the 54 patterns of big data. It's a future-proof approach for building an architecture for the Data era:

- **Data ingest:** Provides the technology to collect, ingest, process, and pre-analyze data. Ingestion is critical because it ensures that the architecture is data-rich, encompassing all possible sources, whether they are internal, external, unknown today, or well known.

- **Analytics:** Where the heavy lifting is done to extract value from data assets. Analytics requires three distinct capabilities: Discovery, Landing, and Operational Warehouse. Each are complementary, yet unique. These are detailed in Chapter 19.

- **Information insight:** This layer is the interface to users, business sponsors, and ultimately business partners and suppliers. It is their view into the data architecture and is the primary consumption vehicle.

- **Operational data:** This represents the traditional data repositories, augmented with new capabilities for entity resolution and data matching. Analytics are only as strong as the operational data area from which they pull much of their structure.

- **Governance:** Incorporating security, along with risk management (disaster recovery, for example), ensures the stability, continuity, and sustainability of the data architecture. It's no longer simply nice to have; it's just as critical as the other components.

Chapter 19 provides detail about each component in the Logical View architecture, along with the key considerations as it evolves to the Physical View and deployment.

18

BUSINESS VIEW REFERENCE ARCHITECTURE

INTRODUCTION

MICHAEL HAMMER WAS an engineer. He earned a B.S., M.S., and Ph.D. at the Massachusetts Institute of Technology (MIT). He later became a professor at his alma mater, focusing in computer science and management. In 1993, Hammer (along with James Champy) wrote *Reengineering the Corporation: A Manifesto for Business Revolution* (Harper Business), which became his life's work, earning him global acclaim and recognition.

The notion of *business-process reengineering*, analyzing and designing workflows and business processes within an organization, was accelerated to a boardroom conversation based on Hammer's book. Hammer's view was revolutionary, as it was based on the premise that most work being done does not create value for customers and therefore should be eliminated, not accelerated. Much of the writings around the topic of business-process reengineering that emerged on the heels of Hammer's work ignored this key premise.

As business-process engineering gained momentum in organizations and with the large consultancies, the focus became automation. Instead of eliminating processes, consultants focused on automating existing processes, often with technology. Advocates of the approach often identified disruptive technologies that could aid in reengineering an organization:

- Shared databases
- Workflow and decision support tools
- Enterprise resource planning tools
- Networking and communication tools

While each of these tools has merit, their introduction to the organization often added, changed, or automated process, but rarely eliminated it. This approach contradicted one of Hammer's seminal beliefs of reengineering: The primary purpose is simplification of process to drive increased efficiency and hence worker productivity.

To achieve this, the practice was focused on breaking down current processes to the smallest possible components and then eliminating those steps that could not be directly linked to customer value. Once the unneeded steps were eliminated, the remaining core components and processes would be streamlined and automated, with the same focus on customer value. Finally, technology could be applied to improve the sharing of information and the execution of the business processes.

The Business View architecture presented in this chapter is not a business-process reengineering exercise. However, it does leverage the techniques. By breaking down the capabilities needed for the Data era into small, well-defined capability areas, a business leader can understand and document the components necessary to win in the Data era.

MEN'S TRUNK: A RETAILER IN THE DATA ERA

All retailers in the Data era face a lot of pressures: margin pressure, customer retention challenges, and typically, an IT fabric that was developed many years ago for transaction processing, long before analytics was possible. For illustrative purposes, and to build off some of the stories provided in Chapter 4, this section will describe a fictional retailer's situation on the cusp of the Data era. The challenges, opportunities, and pressures that are inherent in the business demand a flexible IT environment, to meet current and future needs.

The retailer for this illustration is Men's Trunk, LLC. Founded in the year 2000, Men's Trunk started with an affiliate sales model, attracting young men and women to act as sales consultants. At the time, Men's Trunk had no stores, so all sales were made through their sales consultants, who would host events and parties in local meeting spaces (country clubs, homes, recreation centers, and even local bars and restaurants). Men's Trunk differentiated their business model by being both a merchandiser and a designer. This means that Men's Trunk purchased raw materials from suppliers and turned them into designs that could be marketed through their sales consultants.

During the first five years of operations, Men's Trunk expanded dramatically, as they quickly found young, affluent buyers who enjoyed this type of shopping experience. The key metrics from the first five years of operation are shown in the following table.

Expansion of Men's Trunk from 2000 to 2005

	2000	2005
Sales Consultants	3	72
Customers	10	425
Sales Transactions/Year	70	4250
Revenue	$250,000	$10.6M

While the Men's Trunk story will track the evolution of a new business, the parallels to an existing business are very similar. In the Data era, capturing growth is about reinvention, either by building new businesses or reinventing an existing one.

The Men's Trunk story is just beginning here, and it will continue in Chapters 19 and 20.

THE BUSINESS VIEW REFERENCE ARCHITECTURE

As introduced in Chapter 17, the Business View reference architecture is the key business component building blocks that must be considered for the architecture of the future.

There are six primary components for consideration, captured in the following figure.

ANSWER FABRIC

The *answer fabric* is the set of services that provide insight from raw data. It may be as simple as presenting raw data or as sophisticated as transforming and enriching the raw data to provide a different viewpoint.

The component functionality offered by the answer fabric includes:

- **Visualizing:** Graphically displaying information to a user in a consumable form. This also allows the user to manipulate the views to seek different insights.
- **Predicting:** Executing a predictive model and returning answers of what may happen based on a statistical analysis.
- **Aggregating:** Consolidating information from a variety of sources to feed into an appropriate process for analysis.

- **Organizing:** Organizing data sets, developing ontologies (framework for representing knowledge), and/or creating hierarchies of information.

- **Matching:** Deduplicating similar or identical information, to eliminate duplicate copies and create a merged set of data.

- **Annotating:** Enriching data, either through intelligent tagging of information or augmenting data for new insight.

- **Accessing:** Creating, updating, connecting to, and having the ability to utilize stored information.

- **Alerting:** Interrupting processes or decision-making with notifications of certain conditions or variables.

- **Sensing:** The ability to detect events or conditions that may impact analysis or conclusions.

- **Timing:** Understanding the sequence of events or data by knowing the time dimension of any inputs or changes.

DATA VIRTUALIZATION

Data virtualization acts as the Information Bus for an organization. The bus has a variety of stops throughout the organization. At each stop, the bus collects, collates, and delivers information to where it is required.

The functionality offered by data virtualization includes:

- **Search:** The ability to discover, browse, and explore data, regardless of where it resides.

- **Connectors:** Providing access to information, regardless of where it sits, and allowing for two-way communication and updates. This also provides access to data through known connection protocols.

- **Localizing:** Providing local copies of data for the users or applications that require it. This is important for user experience.

- **Federation:** This makes multiple data sources appear as a single data source, by correlating information and presenting it in a consistent shape.

- **Replication:** Copying and sharing data across different locations, while ensuring that they remain consistent.

DATA ENGINES

Data engines provide the horsepower for managing information.

The functionality offered by data engines includes:

- **Database:** Structured data, typically acting as an operational data store for customer, supplier, transactional, and other key sources of business data.
- **Content repository:** Unstructured data that is typically found in documents, presentations, email, and other natural language type forms. This may also include photos, videos, and audio files.
- **File system:** Grouped storage of information for general use across the engines.
- **Data transformation:** The engine that enables data to move around the organization and to be transformed into consumable form.
- **Operational data hub:** Managed operational information that can be consumed as a point of reference to all other systems. This may be a single source of customer data, for example.
- **Data exploration:** The engine that enables data to be discovered across the organization. This engine indexes multiple sources of information and either pushes that information to other places or can be consumed via search.
- **Metadata engine:** The engine that maintains information about the information. As an analogy, this would be the map of data in an organization; where to find it, how to access it, and how to utilize it.
- **Streaming engine:** The engine that enables high-speed processing of data flows without requiring the data to be stored as a prerequisite.
- **Map-Reduce engine:** An engine for distributed processing of data, leveraging the resources of a large number of systems.
- **Warehouse engine:** The engine that enables analytics of structured data. This engine can take many forms, ranging from a data mart (small, focused), to a sandbox (experimental), to an enterprise data warehouse (all analytics for an organization).

MANAGEMENT

The management layer provides the capabilities for managing the components of the business view architecture.

The functionality for management includes:

- **Install and upgrade:** The capability to add new functionality or augment existing functionality. This may include the addition of new business processes or applications to the data flow or the addition of new data engines. This is required not only to maintain the environment, but also to continue to evolve it as business needs change.
- **Configuration:** This functionality is primarily resident at the sub-component level. It is the ability to configure the components to align with the business processes or applications defined and to ensure that these components are integrated in their interaction.
- **Security:** Security must exist at a variety of levels in the Business View architecture. Security monitoring, data access, network access, and application access are four discrete levels of security required. Each are unique but must be integrated as a comprehensive solution to ensuring security.
- **Backup/Recover:** This is the insurance policy. It offers the capability of backing up key data assets and the overall information architecture, in case of an outage or disaster situation. Quick recovery is an equally important capability to ensure business continuity.

DATA GOVERNANCE

Data governance is the capability to maintain intellectual control over data assets. With the advent of treating data as an asset, data governance plays the role of protecting and managing the life of the asset.

Data governance functionality includes:

- **Protection:** This capability enables the tracking and auditing of all data usage. This may include the ability to mask or anonymize information, as required. Data protection must exist throughout the lifecycle of the asset, until it is retired and no longer accessible.
- **Compliance:** Ensures the implementation of and adherence to key policies. These policies could be defined by the organization or a third-party regulator (like Payment Card Industry (PCI) compliance). Either way, compliance must be monitored, with items flagged that fall out of compliance. Lastly, if something is out of compliance, this functionality must ensure swift remediation.
- **Integrity:** Data integrity is the function that ensures an organization can trust the information that is present in an application, business process, or other consumption engine. As the old phrase goes, "Garbage in, garbage out." Data integrity ensures that garbage does not enter, and if it does, it does not matriculate.
- **Validity:** While this is related to integrity, validity is more focused on ensuring the timeliness and appropriateness of data. It speaks to the quality of the data, ensuring that the right data is used by the right processes.

USER INTERFACE, APPLICATIONS, AND BUSINESS PROCESSES

The user interface, applications, and business processes are the consumption engines. This is what the users will interact and engage with. While the user experience depends on all components, this represents what the user really sees; and it offers a favorable impression, or it doesn't.

SUMMARY

Constructing a Business View architecture does not require any technical nor IT knowledge per se. However, as it evolves into the Logical View architecture, that knowledge becomes more necessary.

The overall objective is to define a Business View architecture, supported by a Logical View reference architecture, to achieve desired outcomes in the Data era.

19

LOGICAL VIEW REFERENCE ARCHITECTURE

INTRODUCTION

THE LOGICAL VIEW architecture provides the building blocks for a Physical View architecture. At the same time, it ensures that an organization is designing with the end in mind. The Logical View architecture is essential to future-proofing the decisions that are made on the journey to a big-data architecture.

The Data era rose to prominence around 2008, with significant experimentation around 2012. The experimental phase, which continues to this day, was originally quite narrowly focused; one use case, or one technology. As we've moved through the experimental phase of technologies like Hadoop (the open source framework for processing, storing, analyzing large amounts of data) and the hype of the whole notion of big data, clients are taking a much more strategic approach to the topic. It's less about trying out the flavor of the month (Cassandra, Mongo, etc.) and more about figuring out how to integrate many of these components into an existing environment in a company. After a few years of experimentation, a reality hits home in medium to large enterprises: They realize that they have to make this work given what they already have in place. It's the classic balancing act between legacy and the future.

A key tenet in developing a strategy for the Data era requires an organization to take a page from Porsche's product strategy and acknowledge that one size does not fit all. There are many technologies, most have a unique and special purpose, and the leaders in the Data era will leverage all or most in a complementary way. For example, Puma realized that the Data era provided an opportunity for the firm to distinguish itself from competitors by quantifying its impact on nature using the Environmental Profit and Loss accounting system.

IT environments of the last 20 years have been largely revolved around traditional data repositories, providing data to business applications. Certainly, as data warehousing and analytics have risen to prominence, we have seen more investment in data-analysis capabilities. However, most of that investment to date has been around augmentation of existing repositories (i.e. providing analytics of structured data from transactional systems). The Data era will require much more than augmentation or incremental improvement. The Logical View of the big-data architecture will demonstrate how the improvements of the past are far too simplistic for the new era.

MEN'S TRUNK: A RETAILER IN THE DATA ERA (CONTINUED)

The exponential growth at Men's Trunk placed early demands on the IT environment to support the operation. Given that the majority of the stakeholders (sales consultants) were not in a single location, Men's Trunk needed a distributed architecture that could support 24/7 access to transactions and information. Between the years 2000 and 2005, Men's Trunk made investments in the follow components:

- **Database A:** Men's Trunk developed an enterprise database to serve as the transaction-processing backbone for the company. All sales orders from consultants were entered in the order management system on the website and recorded into the database. This database contained all history of orders: customer name, address, phone number, items ordered, sizes, date ordered, etc. This became the system of record for the company, capturing the key business transactions.
- **Database B:** This database recorded all transactions with suppliers of Men's Trunk. Every time the company bought different fabric styles, fabric colors, stitching materials, or anything that would go into the clothing, a transaction was recorded in this database. This became another key system of record for the company on the supply side of the business.
- **Data warehouse:** Men's Trunk, around 2003, realized that they needed an easier way to understand customer and supplier behavior in order to

optimize their purchasing and to help the sales consultants through the sales process. They developed a data warehouse, which pulled data from both Database A and Database B, in order to provide this insight.

- **Business intelligence:** The sales consultants and the employees of Men's Trunk required an interface to view the data warehouse. Accordingly, the company acquired a business intelligence tool to analyze the data in the data warehouse. This was an environment in which historical trends could be analyzed and the company could also leverage predicative analytics to model scenarios for the future.
- **Data movement:** As the number of systems expanded, Men's Trunk acquired tools to move data between the various systems and to pre-sort data before it was loaded to the warehouse.

Men's Trunk made tremendous progress in its first five years. The IT environment was built on the fly, as the company expanded. Despite the just-in-time nature of these changes, the IT team was able to keep pace and to provide what each stakeholder required (for the most part).

However, nothing could prepare the technology team at Men's Trunk for what would occur over the next decade. In 2005, Men's Trunk decided to try operating retail stores in order to augment the direct-sales model. In order to not disintermediate the consultants, they made two key strategic decisions:

1. They would sell different (yet complementary) items in the stores than what the sales consultants sold.
2. Each sales consultant would earn commissions from items purchased by their clients in stores.

This revolutionary retail model led to wide praise and acceptance by clients, the fashion press, and the sales consultants. And it set off seven years of tremendous growth. At the same time, Men's Trunk decided to expand internationally, starting operations in over 30 countries. Based on this growth in the business, Men's Trunk looks very different in 2014 than it did in 2005 (see the following table).

Men's Trunk in 2014

Sales Consultants	435
Customers	100,000
Sales Transactions/Year	1M
Revenue	$2.3B
Stores	45

This unprecedented growth has required Men's Trunk to rethink everything they are doing on the technology side to support the business. They must evolve the existing infrastructure, while also providing the capabilities needed for the future.

THE LOGICAL VIEW REFERENCE ARCHITECTURE

The Data era demands a new approach to the Logical View architecture. While previous eras had called for an enterprise data warehouse or other similar one-size-fits-all approaches, it has become apparent that is no longer sufficient. The business demands represented by the Men's Trunk example demonstrate how the silver bullet (one-size-fits-all) approach is not practical.

The Logical View architecture for the Data era is based on five key components, as shown in the following figure. These components are discussed in detail in the rest of this chapter.

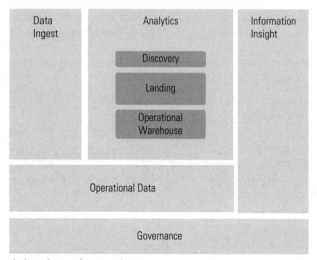

The logical view reference architecture

Depending on the existing environment in any given organization, the starting point may be different. And, irrespective of the existing environment, an organization will evolve to service these components, as opposed to this being a one-time event. That being said, competing in the Data era will require capabilities in each of the five key components: data ingest, analytics, information insight, operational data, and governance. The 54 patterns identified in Chapters 12 and 16 have demonstrated that requirement.

DATA INGEST

Data ingest, to some extent, is the starting point of the big-data reference architecture. Without a strategy for ingest, there is effectively no data to act on.

Data ingest includes all approaches for collecting and performing initial processing on raw data from a multitude of sources. It would include staging areas, along with transformation engines that can perform enrichment and transformation operations on the raw data flows. That being said, there are different levels of maturity around data ingest.

The key capabilities are

- **Streaming engines:** Analyzing data as it is flowing, instead of waiting for it to land in the analytics area. This is a continuous analytics capability, analyzing and parsing data in real-time.
- **Data movement and cleansing:** Moving data from one system to another, while having the ability to understand the shape of the data and provide transformation services as/if required.
- **Data connections and extraction:** Connecting to a vast array of data sources and extracting data in a usable form. This becomes a greater challenge as the Data era drives us toward intelligent machines and other non-traditional sources of data.
- **Data provisioning:** Efficiently assessing the proper data source(s) and driving the data to the right fit-for-purpose processing engine.

Data ingest can provide competitive advantage, as many organizations view this as mere data movement when in fact it is much more substantial.

ANALYTICS

The analytics capability is responsible for the vast amount of data processing necessary in the data era and ultimately becomes the engine that drives the insights to the end users. An analytics architecture for the Data era must include data discovery, data landing, and an operational warehouse. Each capability is tuned to provide a specific outcome: fit-for-purpose. A key premise of the analytics capability is to leverage data landing as the center of the architecture. Data will move into the data landing area as a starting point to later be provisioned to other areas as required by applications or users.

DISCOVERY

This is the place for discovery and deep analytics, primarily of structured data assets, but not limited to that. Do you have large complex analytic queries? Do them here. Do you need high-performance analytics? Do it here. The discovery area is where the high-horsepower massively parallel processing (MPP) analytics will be done. Accordingly, this becomes the core analysis and analytics hub for the organization. This will be the most efficient and cost-effective place for high-performance analytics. Obviously, this requires tight integration with the landing area (discussed in the following section).

The key capabilities in discovery include:

- **Sandboxing:** Experimental environment, which can be provisioned or taken down quickly. Provides the ability to run queries, review results, and then explore related datasets.
- **Predictive modeling and forecasting:** Provides the ability to do native advanced analytics and statistical modeling. Advanced analytics users can build their models in this environment, targeting the relevant data assets provisioned from the landing area. This can form the basis of forecasting provided to users.
- **Deep analytics:** Enables the long-running, complex queries that demand an MPP engine. With this complex-query capability, the workloads that used to take days or hours now take seconds or minutes.

LANDING

This is the place where you land your data in its native form. All data types, sizes, and veracity is accepted and expected. It's the innovation manufacturing floor, and as you begin to harvest your data assets, you can send those refined assets to other zones. So, as previously mentioned, it's the starting point. With Hadoop and a variety of NoSQL (a variety of different database technologies — not only SQL) technologies forming the basis of the landing area, it is the most economical place to start with a large amount of raw data.

The landing area must be cost effective and differentiated by analytics and analysis (not just the run-time). With the data being brought into the landing area, an initial catalog (metadata) can be created to better define the shape of the data. This is essential, as the effectiveness of the other analytics will have some dependence on the landing area.

The key capabilities in landing include:

- **All data:** Data is ingested in raw form. It could be data of any variety: structured, unstructured, internal, external, machine, social, or otherwise.

- **Indexing and search:** Datasets of any size can be properly indexed and then searched as needed. The index built here may become a key aspect of organizing the data for consumption by other areas.

- **Initial discovery:** There is insight in raw data. The landing area enables initial data discovery, perhaps in a sandbox mode, to look for patterns or to understand what else is required.

- **Text analysis and annotation:** The traditional database has necessitated that text be largely ignored for many years. However, much of the data created today (email, documents, messaging, etc.) is mostly text. The landing becomes the place to manage, analyze, and annotate text. Text annotations organize the information for easier consumption.

OPERATIONAL WAREHOUSE

This is the place for mixed analytic workloads. It's not just the high-performance analytics like the discovery area; it can encompass thousands of concurrent users, operational workloads, analytic workloads, and all of them in combination. As the example with Men's Trunk will demonstrate, the combination of analytics and transactional data can lead to unique insights. This capability will be more important in some companies (like credit card companies tracking fraud transactions in real-time) than in others (a manufacturer that produces reports weekly).

The key capabilities in the operational warehouse include:

- **Consolidated and historical views:** With the data shape better understood via the landing area, the operational warehouse can become much more targeted. Consolidated views of data (for example, financial data across an number of business units) can be ingested and analyzed. Extensive historical views of data would be available, as well.

- **Mixed workloads and concurrent access:** Unique insight is gleaned from the combination of transactional data (for example, sales transactions or orders) and analytical data (information about the orders). The operational warehouse enables both types of data to be looked at simultaneously. This area also supports a large number of users seeking the same data at the same time.

- **Trusted data:** Data in the operational warehouse zone will likely have undergone some transformation prior to being loaded. This is necessary because of the types of data being utilized and the corresponding expectation that the users will have for the veracity of the data.

The analytics engines provide processing power when it's needed, for the right type of workloads. Equally valuable is the ability of these engines to stretch back in time, viewing data from a lifecycle perspective. Data is one thing. Understanding data in time solves an entirely different, and often relevant, challenge.

INFORMATION INSIGHT

Information insight is the part of the architecture that most people are naturally familiar with because it is what the user sees. In traditional IT environments, it is the business applications, the reporting and dashboards, and the realm of the initial business intelligence tools. But the Data era demands more.

The information-insight area must enable the consumption of authoritative and insightful information that will improve the operation of the organization. The users have no reason to be aware of all the plumbing behind the scenes, and appropriately so — this should be self-service. Any user, any time.

The key capabilities in information insight include:

- **Applications, user reports, and dashboards:** In many cases, this is the collection of tools an organization has today. They will not and should not be abandoned. Instead, they will become more valuable in the Data era. These will be extended to include new data-focused applications, as well as mobile and web applications that help democratize the access to data.

- **Advanced analytics and data mining:** Statistical modeling, regression testing, simulations, and deep data mining are typical approaches to interrogating datasets. While the tools may change, this capability will be utilized just as prominently in the Data era as it was previously.

- **Collaboration engines:** Insight will become a form of social collaboration. When a user discovers something interesting, the first thing she wants to do is to share it with her boss or a colleague. An engine that enables collaboration and sharing therefore becomes an essential part of the architecture for information insight.

- **Visualization and Exploration:** Perhaps a user does not want to use the data that an application provides. Perhaps the typical business intelligence reports are not sufficient. Or perhaps the user simply does not know what question to ask. Visualization and exploration solves that problem. This capability gives the user the power to comb through and visualize large datasets, looking for patterns or anomalies.

This assumes a solution for data virtualization, which could logically lie somewhere in between information insight and all of the other data sources.

OPERATIONAL DATA

Many of the data assets in an organization are application-independent. But there is still a need to organize that data so that it can be better utilized by applications, the analytics engines, and the information-insight area. This would include, but not be limited to, master data, reference data, metadata, and content repositories. This data is critical to enabling the analytics sources to provide better insight and answers.

The key capabilities in operational data include:

- **Master data:** A single version of the truth. Customer data is stored across hundreds and maybe thousands of repositories. Master data provides the ability to get a single view of that customer, which can become a reference point for other systems. The same applies to product data or essentially any type of entity.
- **Reference data:** Data that is used to categorize or organize information assets, which may include, but not be limited to, master data.
- **Content repositories:** Unstructured content represents the majority of new content being created in organizations today. This could range from presentations to documents, and even include photography or videos. Unstructured content is required to augment a variety of business processes and applications.
- **Activity logs:** The heartbeat of an organization, tracking everything that happens in and around a system. Logs maintain the forensics of what occurred and when. On their own, they could be useless. Combined with other insight or applications, logs often provide a unique insight.

GOVERNANCE

Governance is often relegated to an afterthought. Once an organization has the data they want, then, and only then, do they start to think about governance. In the Data era, that's too late.

Governance is an intentionally broad term, referring to the protection, management, and improvement of information, while demonstrating compliance with key policies, practices, and regulations. Further, governance ensures that access to information is for only authorized purposes. Lastly, governance must ensure that a company can quickly and efficiently recover from any unforeseen circumstances.

The key capabilities for governance include:

- **Lifecycle management:** Data, like all other assets, has a lifecycle. Some assets should be kept forever. Others should be discarded as soon as practical. Data must be viewed like any other asset when it comes to lifecycle management. The ability to manage data and its lineage over time is critical to the governance process.

- **Metadata catalog:** A catalog of data, describing the data assets in an organization. The catalog organizes the data into logical groupings and labels them for easy access and discovery.

- **Data access security:** Whether it's a database, a content repository, or an application, a business must know who is accessing and why. This is more than monitoring, this is understanding who did what and when, which often explains why.

- **Security monitoring:** In a world with an infinite amount of internal and external threats to security, proper governance ensures a way to monitor what is happening across an enterprise architecture. This is complementary to data-access security, as that will typically feed into the monitoring dashboard.

- **Data Privacy:** The ability to mask sensitive data so it can be utilized by other systems, without exposing the sensitive nature of the data.

- **Business continuity:** Business continuity is a step beyond highly available IT infrastructure. This extends the concept to encompass physical security, as well as disaster planning. Business continuity ensures that operations will continue, even in a force majeure situation. This requires strategic planning around multi-site locations.

MEN'S TRUNK: A RETAILER IN THE DATA ERA (CONTINUED)

Men's Trunk grew exponentially in its first five years of operation. However, none of that could have prepared it for the business in 2014, which has evolved to multi-channel and multi-country, and is serving more constituents than it ever has in the past.

While the Men's Trunk example illustrates a company that is relatively young (just 14 years old), the challenges are nearly identical for more established companies. In both cases, the constituents served have changed, either in shape or number or both. The data generated and required is much more varied and complex. And decisions must be made at a faster rate than ever before. Chapter 20 will illustrate how Men's Trunk will evolve their architecture for the Data era.

20

THE ARCHITECTURE
OF THE FUTURE

MEN'S TRUNK: A RETAILER IN
THE DATA ERA (CONTINUED)

MEN'S TRUNK HAS been on a tear since 2005. Their revenue has grown from a mere $10 million to over $2.3 billion. This exponential growth has driven their customer base from less than 1,000 to over 100,000. In line with that, the number of sales transactions per year has grown from about 4,000 to over 1 million. They have expanded into over 30 countries, with 5 new countries expected in 2015. And lastly, the business decision to build physical stores has augmented their web channel with over 45 stores, expected to grow to 62 in the next year.

Recall from Chapter 19 that between the years 2005 and 2014, Men's Trunk began to mature their IT environment to deal with this expansion. Their initial investment included some databases, a data warehouse, data movement tools, and a business intelligence platform to interact with constituents (customers, suppliers, and employees).

The landscape has changed for Men's Trunk, based in part on their tremendous growth, along with their innovative business practices. Here are a number of the initiatives introduced or planned that must be accounted for:

- **Loyalty program:** The company launched a loyalty card program, which is both a physical card and a card stored on a mobile device. This card tracks all customer purchases, gives redeemable points for purchases, offers recommendations, and informs customers about sales and promotions.

- **Multi-channel customer service:** The customer base likes the fact that they can receive chat support online, email support anytime, phone call support most hours of the day, or in-store support during store hours.

- **Personalized shopper:** The company is now offering a personalized shopping experience, where a stylist is assigned to those customers who need help selecting a wardrobe for any occasion.

- **Mobile application:** The mobile application is designed to bridge the web experience with the in-store experience.

The Data era has thrust itself upon Men's Trunk, based primarily on the growth initiatives described earlier. The data environment requirements have evolved to

- **Transactional data:** Given the growth in transactions, this dataset has exploded to nearly 100 terabytes, from a mere 5 terabytes three years ago.

- **Unstructured data:** The company is generating nearly 500 terabytes of unstructured data, primarily composed of social media sharing (pictures, text, etc.), customer feedback (captured from the customer service channel), and location data (geospatial data, based on customers engaging with personal stylists).

- **Application data:** The mobile application is generating new streams of data every day, as customers utilize the application both in and outside of stores.

The chief operating officer of Men's Trunk is the primary recipient of the burden that these initiatives, and the corresponding impacts, place on the company. She sees customer service on the brink of cracking and must prepare the team for the future. She is smart enough to not simply hand a set of requirements to IT, knowing that may miss the mark. Instead, she collaborates with IT to establish a Business View reference architecture, and then asks IT to recommend a Logical View architecture to support that vision. But before jumping to solutions, she urges the team to take a step back and think through the overall implications of the business drivers; this starts with the methodology for applying big-data patterns.

MEN'S TRUNK: APPLYING THE METHODOLOGY

Striving to be a leader in the Data era, Men's Trunk adopts the methodology for transforming their business and architecture. Each of the seven steps bring out critical thinking and consideration for the Men's Trunk team.

STEP 1: UNDERSTAND DATA ASSETS

Men's Trunk has established a team which includes representatives from IT, finance, and the business functions. This diversity provides the span of knowledge required to assess and understand their current data assets. Their existing data is pretty well known within the company. As mentioned in Chapter 19, they have some databases and data warehouses today, largely accessed through the business intelligence tools that were implemented in years past. However, as described in the methodology in Chapter 16, understanding data assets goes well beyond the existing data of an enterprise. Men's Trunk must also understand new data, consider the potential of harnessing external data sources, design an approach for data collection, and clearly document data gaps.

New Data

There is a significant opportunity to analyze new data based on the company initiatives around loyalty programs, personalized shoppers, and their mobile application. Those initiatives, combined with the data collected through the multi-channel customer service, all reflect incremental sources of data. The incremental data from these sources will often be in an unstructured format, particularly that which comes from the personalized shopping and customer service sources. In those cases, the data is often in text or note form, requiring a way to collect and normalize those types of data. For example, sentiment analysis techniques will be required to make sense of the unstructured data and to decide whether or not customers are happy with their services.

Data Gaps

Through the process of understanding their data assets, the team at Men's Trunk also identifies a significant data gap. They realize that their myopic focus on their own customer data and their own stores has ignored relevant external data from fashion houses, along with retail data and relevant trends. This caused an issue last year, when they had excess supply for a particular style. While their customer base had expressed an interest in this style, that interest ignored the fact that the raw materials for that design had increased 50 percent in price in the previous six months. And by the time Men's Trunk delivered the product, the price was so high that the customer base did not engage as expected. This gap in external data is something that has to be closed or narrowed as part of their go-forward planning.

STEP 2: EXPLORE THE DATA

With the data they have today, along with the data that they are starting to collect from new channels, the Men's Trunk team can start to perform meaningful data exploration exercises, probing certain questions, while also searching for new and statistically significant patterns.

Following the methodology, they lay out a set of challenges and questions, leading to the formation of hypotheses (as shown in the following figure). The top graphic shows the flow chart for Men's Trunk of wanting to get from raw data to improved decision-making. The bottom graphic shows how to operationalize this process for addressing a particular high-level challenge that has been posed by management. By converting the question into a falsifiable hypothesis, it is possible to use data to test the hypothesis and to measure the statistical significance of the outcome.

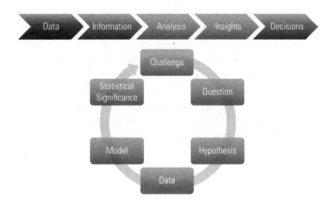

A key hypothesis that the team forms is around purchase frequency. A preliminary analysis suggests that customers who make an initial purchase and then make a follow-up purchase within 30 days generate twice as much revenue over their lifetime as others. Analyzing this hypothesis requires data not only about customers and their purchases, but also the important dimension of time and frequency and how it impacts behavior.

STEP 3: DESIGN THE FUTURE

Design for the future becomes a primary focus for the Men's Trunk team, as their analysis in Steps 1 and 2, combined with the growth and changes in their business, have made it apparent that their current thinking and infrastructure will not support the business requirements going forward. This is explored in depth in the Business View and Logical View for Men's Trunk, covered later in this chapter.

STEP 4: DESIGN A DATA-DRIVEN BUSINESS MODEL

While Men's Trunk has collected extensive and personalized retail, style, and customer information, the company decided that this data is an enabler to the business model, as opposed to something that should be exploited as a business on its own. It's a tough decision, as they know that both designers and large retailers would likely pay a significant premium to have access to their data assets. However, the fear is that they may cannibalize the core business of Men's Trunk.

STEP 5: TRANSFORM BUSINESS PROCESSES FOR THE DATA ERA

Men's Trunk core business processes were formed between 2000 and 2005, when the company was in its earliest growth phases. It was at this time that they created processes for order placement, fulfillment, and other related aspects of the complete demand/supply chain. Although these processes were sufficient for the time, based on the number of customers and transactions, they are woefully out of date today. The primary issue is that the processes have not scaled for the increased business throughput and stakeholders. The team identified three core processes that must be redesigned, comprehending the new stakeholders and new technology platform: supply chain, order processing and management, and procurement.

STEP 6: DESIGN FOR GOVERNANCE AND SECURITY

Men's Trunk recognizes that data is an asset. However, as stewards of personalized information, the company carries an even greater responsibility. In addition, as a retailer collecting credit card information, Men's Trunk is subject to Payment Card Industry (PCI) compliance. To make matters more complicated, Men's Trunk has witnessed a number of retailers over the past few years that have had security lapses, resulting in credit card information and other personalized data being stolen.

Given all of these complexities, Men's Trunk decided to put a significant focus on data governance, privacy, and security. They map out the focus areas as follows:

- **Regulations:** Men's Trunk identifies all local and national regulations, like PCI compliance, that they must comply with. Given the nature of their business, this becomes a significantly long list.

- **Data lifecycle:** The team maps out the lifecycle of data flows, from the point that data is collected to when data is disposed of. This lifecycle must be comprehended in the architecture of the future. Due to regulations like PCI compliance, part of the lifecycle mapping specifies when data must be masked, as it flows through the organization.

- **Data security:** Men's Trunk documents profiles of employees and suppliers, specifying who can access what and when. These profiles are then implemented into a security framework that includes monitoring, along with access and auditable tracking. This is also important to protect customer data in the organization.

- **Physical security:** Given their store presence, security must extend to the physical aspects, as data is collected on location in many stores, representing a different type of potential exposure.

STEP 7: SHARE METRICS AND INCENTIVES

In the early days of the company, the Men's Trunk team always rallied around shared goals. These goals were simple to state and measure, yet hard to achieve. They were goals like new customer acquisition, customer loyalty, and return rate. The Data era will enable a much more granular way to share and measure these metrics, and for directly linking those goals to incentives.

For example, customer loyalty used to measure only repeat purchases. Any transaction in which the customer bought a second time was deemed to be a success, even if he bought only that one additional time. They were also deemed loyal if they bought frequently, even if they spent less money than the previous year. The metrics were limited by the data collected and utilized.

With the new data architecture, Men's Trunk can now transform measuring loyalty so that it's more tied to customer engagement. Customer engagement will be measured by the multi-channel interaction (web, mobile, store, or all three), purchases made based on recommendations from personal shoppers, and customer's recommendations (if they recommend Men's Trunk to others, via social media or otherwise). New data sources, augmented with the ability to analyze and combine those data sources, puts an entirely new lens on metrics. With more granular and relevant metrics, building incentives becomes quite easy.

MEN'S TRUNK: THE BUSINESS VIEW REFERENCE ARCHITECTURE

Men's Trunk recognizes that a cross-functional team will bring the skills required to define a sustainable Business View architecture. This team includes representatives from retail, merchandising, and the supply chain. In addition, the team will include operations experts (from the COO's team), along with IT infrastructure. The team quickly agrees that the framework shown in the following figure represents the key considerations for building out the requirements.

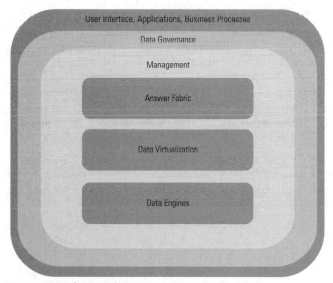

Business View Reference Architecture

They also decide that they will not need to comprehend every sub-component in each category and agree to focus on the priority items for the next five years.

ANSWER FABRIC

The answer fabric of Men's Trunk will provide the insight demanded by the key constituents: employees, suppliers, and customers. The team agrees to prioritize:

- **Visualize:** A user interface, unique for employees, suppliers, and customers. In the case of customers, they will be able to access data about their historical purchases and receive recommendations based on what other similar customers have purchased. In the case of suppliers, they will have access to a real-time view of inventory (for replenishment), as well as a view of their receivables and payments. Employees will receive a single view of each customer, regardless of the channel (store, web, or mobile application). Employees will have access to data that explores style, store, and seasonal trends, as well as insight into customer behavior.

- **Predict:** Men's Trunk requires a predictive capability to forecast sales of new product introductions. In the past, not having a keen sense of what styles would catch on, the company ended up with too much inventory or a shortage. Their new product introduction process will now incorporate predictive capabilities based on a variety of factors to lessen the chances of a shortage or surplus.

- **Aggregate, organize, and annotate:** With data now coming from over 100-plus systems — across stores, the web, mobile applications, and different data types — Men's Trunk will aggregate and organize these data flows. They will start building hierarchies of the information, collecting metadata for easy access later. They will enrich the data collected by analyzing the text of the customer correspondence and bringing in relevant third-party data.

- **Access:** To accomplish many of the tasks described thus far, data access and consistent updating is mandatory. Each constituent must have the ability to utilize current information.

- **Alerts:** Alerts will be initiated from the sensing application, notifying constituents of a changing condition and the action required. A big focus for Men's Trunk is on customer returns. They have noticed a trend where customers will buy online and then return in stores if the product is the wrong size or style. When a customer initiates a return online and specifies that he will come into a store, Men's Trunk wants to be prepared with an individually styled portfolio to present to the client when he arrives.

- **Sensing:** Tied to the predictive capability, to sense and act on any shortage or surplus in new styles. This capability will also ensure an ongoing adaptation to a change in preferences. As customer preference changes, suppliers, store displays, and web/mobile content should be updated to reflect the evolving preferences.

DATA VIRTUALIZATION

Data virtualization will ensure that the right data is in the right place at the right time for each of the constituents of Men's Trunk.

The team prioritizes:

- **Search:** Employees will have the ability to explore preferences and patterns, to create personalized recommendations for customers who have made an appointment in a store. While some of this will be done with the recommendation engine, Men's Trunk has found that the engine, combined with an employee's sense of style, leads to the best results.
- **Connectors and federation:** With over 100 data systems at this point, the infrastructure must be able to easily connect and federate data from each of those sources. The company has found that customers often shop in multiple stores, as well as online, and that data must be federated when an employee is interested in a certain customer's preference or history.
- **Replication:** Primarily used to copy data and share it with the disaster recovery site that Men's Trunk has created.

DATA ENGINES

Given the new scope of data collected by Men's Trunk, the team envisions significant growth to their data engines. These engines provide the horsepower for managing the information:

- **Database:** Each store will have a database, collecting transactional data from each purchase. In addition, the mobile application has a database that is mirrored across a number of geographical sites. There are also databases housing the supplier data. In short, Men's Trunk has seen their infrastructure explode, with nearly 100 databases to support data collection and transaction processing.
- **Content repository:** Will likely include the multitude of photos and videos of new fashions and trends that Men's Trunk catalogs on an ongoing basis.
- **Operational data hub:** Will house the data of record, to serve other systems. Men's Trunk will need to create a single view of their customer records to effectively operate customer loyalty and to provide actionable analytics.
- **Data exploration:** Part of the design point in this transformation for Men's Trunk is to democratize access to data. Each of the stakeholders in Men's Trunk need enhanced data to perform their roles.

- **Streaming engine:** The engine that enables high-speed processing and ingestion of data from retail stores and other sources, without requiring the data to be stored as a prerequisite.
- **Map-Reduce engine:** An engine for distributed processing of data, leveraging the resources of a large number of systems; at the center of Men's Warehouse's data collection and analytics strategy.
- **Warehouse engine:** The engine that enables analytics of historical data, based on customers, products, and suppliers.

MANAGEMENT

Men's Trunk has always had a strong discipline around managing their environment and introducing changes to the production workloads. Therefore, most of their attention for Management falls to security and backup and recovery needs:

- **Security:** As previously mentioned, security monitoring, data access, network access, and application access are four discrete levels of security required. While each of these levels is important, the initial focus for Men's Trunk will lie in data access and monitoring. Given the sensitivity of the data collected, Men's Trunk must harden their approach to data security as a first step.
- **Backup/recovery:** As Men's Trunk establishes their Data-era architecture, the next step will be to set up a different physical site for remote backup, to protect against any risks to business continuity.

DATA GOVERNANCE

Data governance is the ability to maintain intellectual control over data assets. With the advent of treating data as an asset, data governance plays the role of protecting and managing the life of the asset.

Data governance functionality includes:

- **Protection:** Given the nature of the data collected by Men's Trunk, tracking and auditing of all data usage is critical. They will need the ability to mask or anonymize information, so that personally identifiable data is not shared.
- **Compliance:** Compliance ensures the implementation of and adherence to key policies. For an offline-and-online retailer like Men's Trunk, which collects credit card information and other sensitive data, there are a number of compliance regulations to be considered.

- **Data validity and integrity:** Trusted information is required by each of Men's Trunk stakeholders. They are using the data to make purchasing, as well as recommendation, decisions.

USER INTERFACE, APPLICATIONS, AND BUSINESS PROCESSES

Men's Trunk is a social company. It's often referred to as a family, more than a company. That comes out in the employee relationships, along with the customer engagement. Many Men's Trunk customers know some of the employees (personal shoppers in stores or even online), and that relationship is part of the experience.

A key design point around the applications for data access is to facilitate that social experience. Therefore, a key requirement is to leverage capabilities that enable sharing and collaboration between employees and customers to be sure that the personal touch is not lost:

- **Business processes:** As previously mentioned, Men's Trunk redesigns three key processes: supply chain, order processing and management, and procurement. Each of the processes and workflows will be manifested through a rich set of web-based applications so that they can be accessed easily from anywhere.

- **Data science and visualization:** A special part of the company culture is the engagement between employees and customers. Men's Trunk will utilize data-visualization tools to enable employees to better understand customers, behavior, patterns, and trends. All of this will enhance the engagement. And, per the social design points, the data is easily shareable.

- **Customer engagement dashboard:** Customer loyalty is everything for Men's Trunk. Therefore, the company will have a customer loyalty dashboard to look at loyalty and engagement on a personalized basis and on a segment/location basis.

- **Mobile adviser and shopper:** Customers have often said that they wish they had a Men's Trunk employee with them on a shopping trip, even if it is not at a Men's Trunk store. The company has built this trust through their engagement culture. The mobile experience becomes about not only shopping with Men's Trunk, but also helping and advising customers anywhere they are shopping. An advisory role builds trust, which leads to higher customer engagement and loyalty over time.

MEN'S TRUNK: THE LOGICAL VIEW REFERENCE ARCHITECTURE

With the work around the Business View architecture well defined, the Men's Trunk team must now begin to operationalize it through the Logical View architecture. The purpose of the Logical View architecture is to take the capabilities from the Business View architecture and to decide how those will be executed in the architecture of the future.

The Men's Trunk Logical View architecture touches on all of the major components of the architecture shared in Chapter 19. Given the requirements established in the Business View, Men's Trunk will require significant investment in analytics and operational data, as those components will serve as the backbone of the architecture.

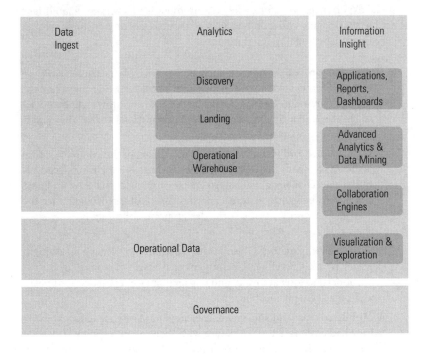

APPROACH

Men's Trunk, a consummate innovator, leverages the Logical View architecture, but with its own unique approach.

The Men's Trunk team, comprehending the significant growth in data and data sources, along with the varied types of data, decided to place the analytics landing area at the center of its Logical View architecture. They select Hadoop as the technology for the landing area, seeking to exploit a number of attributes:

- **All data:** All data will be landed here to start, in its native form. It will include application, mobile, transaction, text, social, and many other types of data.

- **Indexing and search:** As the large datasets are collected in Hadoop, they will be indexed and organized through annotation and the creation of ontological models. From this point, the data in Hadoop or outside of Hadoop can be indexed and searched. The landing area becomes a lens to all data in the organization.

- **Text analysis and annotation:** For the unstructured data ingested in Hadoop, the landing area provides the capability to manage, analyze, and annotate text. This will likely include data from the multi-channel customer service platform, and perhaps inputs from personalized shoppers.

With all data landed in Hadoop as a starting point, Men's Trunk can provision data to other engines to perform special-purpose functions. For example, they will provision their shopper-history data to the Discovery area for high-performance analytics and recommendations. They will provision much of their transactional data, combined with some of the behavioral data, to the Operational Warehouse, to analyze those two datasets together. Other data will be accessed directly in Hadoop via the Data Visualization and Exploration interface.

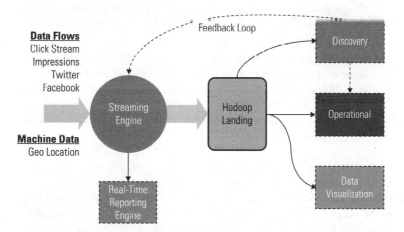

For Men's Trunk, the data ingest capability that they have designed into their architecture enables the Hadoop landing area to exist and constantly evolve as the data assets of the organization evolve. Men's Trunk elects to implement a streaming architecture for data movement, as they want to have the ability to generate real-time reports, based on customer or supplier behavior. The ingest capabilities that they focus on are:

- **Streaming engine:** At the core of their data ingest strategy, providing a continuous analytics capability, analyzing and parsing data about purchases and customer behavior in real-time.

- **Data movement, cleansing, and connectors:** These capabilities enable Men's Trunk to move data from one system to another, while connecting to its now disparate sources of data.

- **Data provisioning:** As mentioned earlier, with Hadoop at the center, data provisioning is an essential function to efficiently assess the proper data source(s) and drive that data to the right fit-for-purpose processing engine.

Men's Trunk decides to invest in an Operational Data layer, to serve as a point of consistency and a trusted source, to all systems and applications.

Men's Trunk scopes their primary needs as:

- **Master data:** A single version of customer records, matched and culled from the disparate systems and customer touch points. This becomes the backbone for their personalization efforts and drives the insight for the loyalty program.

- **Content repositories:** Men's Trunk has amassed a varied and constantly evolving set of unstructured data assets. These assets are primarily photos and videos of styles, which can be viewed and shared as part of the mobile application. The personal shoppers make extensive use of this data while they work with customers.

- **Activity logs:** Serve as a beacon of customer behavior and interaction because the company can use these logs to get a sense of how the systems are being used. For example, in the past six months, the Men's Trunk team feels that a large portion of the shopper browsing has shifted from the web to mobile applications. The logs will confirm or deny this assertion.

With this expansion in data practices and the variety of access points and capabilities, data governance becomes essential. Given that much of the company's competitive advantage comes from the personalized knowledge of their customers and behavior, Men's Trunk places a premium on data access and security. On top of this, the company decides to invest in an advanced security monitoring system, with close attention to monitoring all application access and usage. As these capabilities mature, the company will plan a more extensive investment in data lifecycle management. However, the initial focus will be on data security and system monitoring.

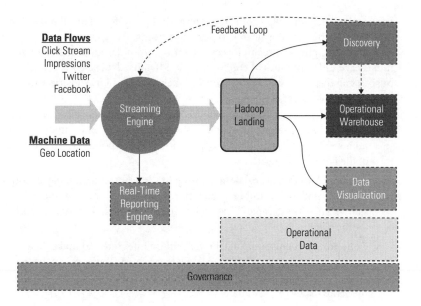

Business continuity will be a priority from the initial launch, with the company planning two nearly identical sites, ensuring business continuity and protection from any unforeseen events.

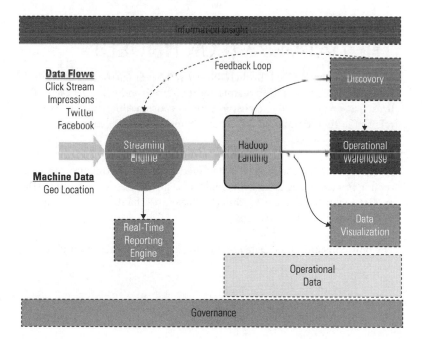

Information insight will become the interaction point for Men's Trunks data investment. The company wants to provide a point of engagement for customers, suppliers, and employees. At a more granular level, the company must serve different types of employees: statisticians, data scientists, and business analysts and executives. Those stakeholders will use a variety of methods to access data:

- **Applications, user reports, and dashboards:** The primary interface points will be web and mobile applications. These points will serve each of the constituent users.

- **Advanced analytics and data mining:** Statistical modeling, regression testing, simulations, and deep data mining tools, all aimed at the statisticians. These capabilities will be used for predictive analysis and testing.

- **Collaboration engines:** Social tools will be integrated to make data analysis a collaborative experience. Data scientists, business analysts, and statisticians can easily share their findings and collaborate around the same datasets.

- **Visualization and exploration:** This will enable direct access to the landing area, primarily to be used by data scientists. It provides access to a rich set of data for exploration.

MEN'S TRUNK: A RETAILER IN THE DATA ERA (CONTINUED)

Men's Trunk has evolved through the years, as their customer's behavior has changed. Recognizing the dynamic nature of their industry, the Men's Trunk team has seen the need to transform and because of this, they have remained relevant to their customer base. These characteristics are common among the leaders of the Data era.

Men's Trunk, adopting the seven-step methodology, continually reinvents their approach to engaging with employees, suppliers, and customers. The application of the reference architectures ensures that they are able to do what their stakeholders require.

The only constant in the Data era is change. Men's Trunk is ready.

Epilogue

THE TIME IS NOW

As with all major decisions in life, it is tempting to sit back and wait for a clear path forward to present itself. Similarly, business leaders may wish to wait for a clear trend to emerge before making a transition towards using big-data approaches within their organization. Some people believe that big data is just a hot topic that will eventually fade away and that continuing with business as usual is a sustainable solution. Big data is indeed a buzzword that means different things to different people. For many, merely processing large volumes of data is sufficient to claim that they are dealing with big data. After all, few would admit to working with small data for fear of appearing unsophisticated. However, processing data is not sufficient. The holy grail of the big data revolution is to construct a seamless framework for ensuring that leaders of business and government are able to consume this data and use it for making decisions and informing policy in a truly knowledge-based society.

Big data strengthens an existing trend in the computerization of every aspect of our lives, as witnessed through the automation of many of the tasks that have been traditionally performed by humans. Automated systems have already replaced receptionists and are now routinely used to answer telephones. By extracting basic information from us, they can direct us to an appropriate department. The response, "The computer says no," coming from service providers highlights just how decision-making is being taken out of the hands of humans and given to machines. This competition between human and machine is everywhere, and the best strategy for survival is to join the revolution — or better still, lead the revolution. Attempting to compete against organizations whose strategies rely on sophisticated machine learning algorithms is likely to be a fruitless struggle.

The true competitive advantage will come from those that manage to overcome the challenges of volume, velocity, and variety — the three Vs of big data — when harnessing big data for decision-making. *Volume* refers to the fact that more data (SMS, email, voice, video, social media, etc.) is being created than can be stored and that sophisticated computational techniques are required for processing this vast quantity of data. *Velocity* is a challenge as we attempt to cope with high-frequency data that require data-stream algorithms to transform this flow of data into real-time decision support to enable organization to respond immediately to threats and opportunities. *Variety* is a key component for describing big data and highlights the need to consider both structured and unstructured sources of data, ranging from official business and government statistics to satellite imaging to social media channels such as Twitter and Facebook. Taking on all three challenges and creating a framework for decision-making is a considerable undertaking and requires a multidisciplinary team of experts, ranging from computer scientists to engineers to statisticians to business leaders.

Another key component regarding the timing of the big data revolution is the ability to make innovative services available via mobile applications. With mobile penetration increasing at a phenomenal rate, many people now use their mobile devices to assist with many everyday tasks. According to the International Telecommunications Union (ITU), mobile penetration rates are almost equal to the global population, at 95.5 percent worldwide, including a 90.2-percent penetration rate in developing countries, which currently have the highest mobile growth rates. Mobile devices are generating big data about human activity, preferences, and behavior, and they will be crucial for delivering decision support based on sophisticated algorithms.

TAKING ACTION

The first step to participating in the big data revolution is to take stock of how one's organization currently approaches the process of planning, strategizing, and making decisions. How much of this process is data driven and how much is based on the hope that the status quo will persevere? Unfortunately, there are numerous reasons for not taking action. Mottos such as "If it's not broken, then don't fix it" warn against change. If a firm has upward-trending sales figures, it is easy to presume that these will persist and that it is best to simply go with the flow. History is littered with stories of organizations that failed to take action despite it being glaringly obvious that their traditional business model was under threat from a new technology. Despite inventing digital photography, Kodak's management ironically failed to appreciate the importance of this disruptive technology, and hesitance to respond saw digital photography eventually destroy its film-based business model.

Maintaining a forward-looking risk register and tracking key performance indicators (KPIs) over time is crucial to monitoring the health of any organization. Just as doctors intervene when our bodies show warning signs of the onset of medical disorders, it is important to continuously check the pulse of an organization. Imagine trying to drive a car without a speedometer and how hard it would be to avoid speeding fines. Worse still, consider how dangerous it would be to drive with a blindfold and no sense of what was coming except for sporadic instructions from a backseat driver. While many organizations now view data as facilitating a vision of the future and providing the ability to respond to dangers quickly, there will be others that prefer to keep the blindfold on. Unfortunately for the latter, it will be difficult to survive when faced with competitors that are equipped with superior information, intelligence, and insight. Big data and quantitative modeling can provide a dashboard for selected KPIs, and it is important that senior management use this dashboard when steering their firms into the future.

The costs of implementing a big data strategy should be quantified and compared with the rewards over a sufficiently long timescale. Making the transition will be disruptive and costly in terms of the necessary resources, but it is important to appreciate that the rewards will flow for a long time. It is arguably easier for entrepreneurs to take such a strategy on board because they will not have to deal with outdated legacy systems and resistance from those who are simply adverse to change due to the personal costs that they fear. The push for change should come from senior management and can also be formulated as a part of a strategy for risk management. In situations in which it is difficult to quantify future rewards, focusing on risk reduction may be the best motivation for change.

FEAR NOT USUAL COMPETITORS

In many walks of life, it is remarkably easy to be lulled into a false sense of security by viewing traditional competitors as the only threat that should be closely watched. The Internet revolution has already shown how quickly new firms have established themselves and managed to compete aggressively against incumbents. Just as the Internet lowered the barriers to entry and allowed entrepreneurs to pilot novel business ideas, we believe that big data will facilitate similar levels of innovation. Perhaps the most important trend that was set in motion by the Internet was an ability to scale up a successful business concept by quickly and efficiently reaching a large volume of subscribers and potential customers. Indeed, in some cases (such as Skype, Twitter, and Facebook), obtaining a loyal group of subscribers was more important than actually transforming that community into a revenue stream. As interesting ideas go viral on the Internet, the cost of marketing is dramatically decreased. In addition, novel approaches to raising seed capital allow potential customers to get directly involved in funding businesses.

Data has always been at the heart of successful Internet firms. Google figured out how to search through vast amounts of data and information. eBay made it extremely easy to buy and sell, and it provides data about how much we can trust other members of its community. Amazon uses data about our previous purchases to construct a list of desirable items that we are likely to want to buy. In short, an ability to understand customer behavior has been at the forefront of the Internet. Big data is not only a commodity for enabling this understanding, but it is also fast becoming the intellectual property that will distinguish between the winners and losers in this competitive marketplace. Raw data has no value unless it is being consumed, and in the same way, intellectual property is only valuable if you put it to use, create patents, and design innovative products.

Big data provides information about who we are, where we are, what we are interested in, what we want to do, and what we are willing to spend money on. Numerous disruptive business models are already arriving that challenge traditional firms and modes of operation. It is safe to say that few of those under threat today could have predicted where the challengers would have come from. Fast, nimble tech-savvy firms that are crunching data and utilizing mobile apps are challenging organizations that are traditionally conservative and adverse to technology due to regulation and legacy business models that allow for slow-paced sales cycles.

Uber is described as a ridesharing service with the motto, "Request, ride, and pay via your mobile phone." It has undermined taxi services worldwide by providing cheaper travel services and flexible employment for many drivers. Since being founded in 2009, Uber has reached 200 cities across 45 countries and was valued at $18.2 billion in June 2014. Airbnb is a website for renting out lodging. It has tapped into the underutilized resource of spare rooms in private houses and helped many to secure additional income by renting out their bedrooms. Founded in 2008, Airbnb was valued at $10 billion in April 2014. Both Uber and Airbnb are innovative in their approaches to unlocking unused resources and utilizing big data to measure trust.

THE FUTURE

Numbers, statistics, and many of the technical aspects of using big data are not likely to be for everyone. But there are definite dangers to hoping that changes caused by the big data revolution will not affect the employment landscape as we know it today. Those that can manage big data and convert it into insight, intelligence, and strategy for leading organizations will be sought after in the knowledge-based society.

As big data provides a means of constructing automated systems for measuring trust and quantifying risk, it has been possible to cut out numerous middle people who previously served to assess individuals and services, and provide quality assurance. It remains to be seen whether big data can also offer a replacement for official licenses, certificates, and standards provided by regulatory authorities. Again, this question will eventually be settled by the cost and accuracy of these approaches to measuring reliability and quality. Based on current trends and both the flexibility and scalability offered by data-driven methods, our bet is that big-data approaches to establishing levels of trust and managing risk will eventually win out.

The fear of being pushed outside comfort zones will not sit well with many employees, but the long-term benefits and rewards offered by the big-data revolution are too large to expect that it will fade away without impacting society. We have offered stories to explain how change is already taking place in various cutting edge organizations. This led us to select key messages about how big data is transforming industry and society. To encourage action, we proposed a roadmap for creating change and using big data within different organizations. We welcome you to join us in the big data revolution and hope that this book will inspire you to take action today.

Index